# HOMESTEADING
## IN THE
# Last Best West

Based on JB Hansen's Personal Memoir

BY
ELAINE MELBY AYRE

# FriesenPress

One Printers Way
Altona, MB R0G0B0,
Canada

www.friesenpress.com

**Copyright © 2022 by Elaine Melby Ayre**
First Edition — 2022

Dedicated to the fifth generation of John and Dina Hansen's descendants

Also by the author: *The Princess Doll's Scrapbook*

All rights reserved.

No part of this publication may be reproduced in any form, or by any means, electronic or mechanical, including photocopying, recording, or any information browsing, storage, or retrieval system, without permission in writing from FriesenPress.

ISBN
978-1-5255-0698-7 (Hardcover)
978-1-5255-0699-4 (Paperback)
978-1-5255-0700-7 (eBook)

1. BIOGRAPHY & AUTOBIOGRAPHY, HISTORICAL

Distributed to the trade by The Ingram Book Company

Soli Deo Gloria

Not to us, LORD,

not to us,

but to your name

be given glory

on account of

Your gracious love and faithfulness

Psalm 115:1

Dedicated to
the
Descendants
Of
John and Dina Hansen

# FAMILY TREE CHART

Here are the birth and death dates for the Hansen parents and children.

John B Hansen– April 7, 1877 – January 27, 1965
Dina Amelia Gronvold– October 2, 1887 – March 29, 1968

1. Edith Marie– March 3, 1910 – October 15, 2008
2. Myrtle Johanna– October 31, 1911 – September 1, 2005
3. Bernhard Rudolph– October 30, 1913 – April 30, 2002
4. Clarence Alvin– May 8, 1915 – October 6, 2011
5. Anna Irene – July 24, 1917 – September 2, 1980
6. Clara Lillian– June 18, 1919 – December 9, 2002
7. Evelyn Violet – May 21 (or 26), 1921 – October 22, 1922
8. Johnny Henry – February 11, 1923 – April 6, 2018
9. Evelyn Alida – October 25, 1925 – June 29, 2012
10. Lloyd Walter– October 13, 1926 – October 22, 2015
11. Palmer Donald– February 21, 1929

# Foreword

"Winning" a pair of baby booties marked the beginning of my link to my grandfather, JB Hansen's homesteading story. It happened like this. When my Grandmother, Dina, died in the spring of 1968, her only remaining piece of finished handiwork was a pair of knit baby booties set aside to be given to her first great grandchild.

Four years later, my cousin, Vivian Tangjerd and I, who both married in 1970, were in contention for the "coveted" booties. Vivian's due date was a week before mine; it seemed like she would be the beneficiary.

But as it came about, I 'won' the booties. My son, William was born shortly after midnight on Thanksgiving Day, October 10, 1972 and Vivian and Warren's daughter, Andrea, made her appearance at the other end of the very same day. Large white booties with a blue band would never have fit the petite Andrea.

William and Andrea, the first two of the JB Hansen family's third generation, born on the same day, reminded us of Pebbles and Bam-Bam from the Flintstones. Don't you think so?

*William Ayre and Andrea Welling- the first of the great grandchildren of the JB Hansen clan. This book is dedicated to the children of that generation.*

***

Now, how did this story connect to the birth of John and Dina Hansen's first two great grandchildren- my son, William Ayre and my cousin's daughter, Andrea Welling?

At the end October 1972, the two new grandmothers returned to Saskatchewan after a stay to help their Alberta daughters adjust to their first couple weeks of motherhood. While Uncle Harold drove, my mother, Edith and her sister, my aunt Clara, sat in the back seat with an old ledger book, handwritten in Norwegian between them. It was my grandfather, John B. Hansen's personal memoir. In the ten-hour trip back to

Regina, Saskatchewan from southern Alberta, the sisters enjoyed translating my grandfather's manuscript from Norwegian into English. Since Grandpa often mixed English and Norwegian expressions, they found it quite amusing.

Clara wanted to have the story printed, but without support to do this, she wrote it out in longhand. Then she photocopied those 82 pages along with 12 pages of pictures to produce a small booklet: **From Skoger, Norway to Hoffer, Sask. -J.B. Hansen's Story.** She gave it to family members for Christmas in 1992, 29 years ago.

My first book, **The Princess Doll's Scrapbook- Her Families' Emigration/Immigration Stories,** alluded to portions of this homesteading story, but it didn't belong there; it was its own story. So for this story, I used that booklet of Grandfather's memoir as the starting point. By combining old family letters, historical tidbits, and assorted memories to add interest and colour and fill in many details, like bits and pieces in a crazy quilt, here is **J.B. Hansen's Homesteading in the Last Best West,** thanks to Clara and Edith's translation efforts, and Clara's determination to record the story, so we would have it.

***

Initially, I anticipated completing this book as a commemoration of Canada's 150th birthday when I submitted a very rough version of this manuscript to FriesenPress to begin this book's publication journey in February of 2017. Then it had been fifty years before, in the 1967 Centennial year, when I lived with my grandmother while I worked as a home economics teacher at the Estevan Collegiate. Those two years provided a special connection with my grandmother and inspired my previous book: **The Princess Doll's Scrapbook- Her Families' Emigration/Immigration Stories.**

Memories of 1967 Centennial projects pale when I think of the journey Grandmother's gift in the old shoebox initiated. Initially, an eight page "Doll's Story" shared this historical treasure with my family and friends. And when more of grandmother's family history was uncovered, it became

**The Princess Doll's Scrapbook-Her Families' Emigration/Immigration Stories.** That 'Doll Story' ended with JB Hansen's homestead, begun in 1909, with a quick fast-forward to the 2009 celebration of that 100-year-old Farm. Now, it is time for my grandfather's story to be told.

\*\*\*

The fact that I should record and enlarge J.B. Hansen's Homesteading Story was an early January 2017 after midnight epiphany- the burning realization that this account of my Grandparents' homesteading years in southern Saskatchewan needed to be documented to be part of the year we celebrate 150 years of Canada's history. Yes, I know! I missed it by four years.

When I first considered the concept for this second book in 2013, I purchased a second self-publication package from Friesen Press, the publisher of my first book. Five years slipped by; the package needed upgrading and completion, with a manuscript produced post haste, before losing my investment altogether. However, the kind of mashup manuscript I envisioned still required a lot of work to get it right. And then in mid-June of 2018, when I announced I was very close to being finished, a major health issue requiring abdominal surgery put this all on the back burner. Then, after sixteen months of hospital visits several times a week for bandage changes, everything's back to almost normal.

Getting back into the groove was difficult. But the project had to be completed. Then, in February of 2020, it was my husband's turn with a cancer diagnosis requiring 12 weeks of Chemotherapy and July and August spent in Edmonton to get daily Radiation therapy at the Cancer Centre.

And as I was doing this, I recognized the unique position I held between the generations. I first became involved in the story Grandmother's gift inspired. That Doll Story and now this Homesteading Story shares that heritage.

Whether you are connected to this story by family, friendship or just an interest in the accounts of how it was way back, when the province of Saskatchewan was just beginning its modern history, it is my desire and prayer that this book will help you understand and appreciate our pioneer heritage.

***

Let us not the ancestors forget

under everything that turns and changes,

because they left us an inheritance to save,

greater than many can realize.


Let it be remarked, more than in mere words

that we keep that inheritance in good condition,

so that when our descendants again are seen on earth

they can recognize their people and their country.

Ivar Aasen

Ivar Aasen (1813-1896) was a Norwegian philogolist, lexicographer, playwright, and poet. His poem was used as the introduction to the family history book: (Broder Jordbrek, ***Den Gamle Skogerslekten Fremtidens trykkeri,*** Drammen, page 15) and introduces the importance of descendants knowing about their ancestors, as we're doing in the first chapter and this entire book, for that matter.

## Chapter 1
# NORWEGIAN ROOTS

*Beginnings in Norway*

It was their trip of a lifetime. By taking off the school year from their teaching jobs, my aunts, Myrtle and Evelyn completed a general tour of Europe, and Norway in particular, to get acquainted with their relatives there. They left in the fall of 1956 and returned by the summer of 1957.

And prior to this, my grandfather, JB Hansen, gathered the family together on the May long weekend to have a studio portrait taken to present as gifts for his family in Norway.

About six years after this, my great uncle, Broder Jordbrek (grandfather's brother) took over the editing of a family history book called **Den Gamle Skogerslekten** (The Old Skoger Families). We simply referred to it as "The Blue Book." We couldn't read it, for it was written in Norwegian. If someone marked it, we could find the pages where our ancestors were recorded and see the names and birthdates of our great grandparents and their assorted siblings- our distant cousins from the past. Now that book, recently translated[1] by a distant cousin who now lives at Jordbrek, provided this picture of the place in Norway where my grandfather JB Hansen grew up. I thank him for allowing me to use his translation to begin this story.

---

1     Hans Anders Skjeldrum Elgvang, **The Chronicles of The Ancient Family of Skoger** Translation of **Den Gamle Skogerslekton** from Norwegian to English. 2009

# Family from Jordbrek

In my great uncle Broder's words, with mine in brackets to explain, we begin using the translated portion of the text describing Jordbrek Farm, where my grandfather was born and raised:

*"The farm [named], Jordbrek[2], in [the kommune or municipality of] western Skoger, is situated between Skallestad and Svingen [on the border of the fylke [or county] of Buskerud and Vestfold, two districts west of Oslo. Skoger was also the name of the parish.]*

*It is farm number 64 and the farm unit, numbers 1 to 8. [Its] name indicates that it is a very ancient settlement, which was cultivated during the second era of cultivation in our country, because the name is clearly derived from the natural shape of the farm's location. The name Jordbrek probably stems from the parts- 'jord' meaning earth, soil or ground and 'brekk' meaning edge or angle. It is situated upon the tallest spot of a moraine from the Ice Age.*

*In 1348, the name was spelled Jordbrekka, which clearly indicates that the farm is situated high up on a hill. One other farm in Norway of the same name is in Suldal in Ryfylke and it shares the characteristic location.* [Variations in spelling reflect changes that took place in the language in the period after the Black Death, when Denmark took a leadership role in Norway, strongly influencing the language after this period. It is interesting that names of farms and other places were often descriptive of the locale.]

*Jordbrek is mentioned in the* **Drammenshistorie** *by Tord Petersen, together with the first freeholder of the farm, Kristoffer Rasmussen, one of the first we know of from our shared family. He kept the farm from 1634, until his death in 1689. The farm has been in the possession of the family since then.*

---

[2] Remember the "j" sounds like "y" as in the word fjord

*In 1755, the farm was divided into two parts of the same size, resulting in about 800 acres of cultivated land and 1600 acres of forest to each farm. The one farm which goes under the name, Northern Jordbrek, is farm unit number 1, owned by the brothers, Martin and Christian Skjeldrum. The other farm, which goes under the name Southern Jordbrek has the farm unit number 6, and is currently owned by Hans Skjeldrum.*

*Buildings of the farm are about 200 meters apart. In olden days, the country lane passed by Jordbrek, as was the usual custom.*

*Wideroe aerial photo of Jordbrek gard (farm)*
*Photo credit: "Lands Museum" (Randsfjordmuseet)*

*From my childhood – around the turn-of-the-century* (1900), *I so well remember traces of the lane that had passed through the gardens of southern Jordbrek, only a few metres behind the main building, through the longitudinal direction of the orchard and further down towards Skallestad. It was obvious to see that the planting of the old garden had taken place in two separate eras.*

*The first planting was probably in connection with the construction of the main building of southern Jordbrek. This was built in 1832, to judge by a decorative name plate that was put in front of the main entrance. Here the name Jordbrek, written with the digits 18 before and 32 after, was all painted and decorated in a very artistic manner.*

Having your picture taken here is on the bucket list of many of the descendants from this place.

*My uncle, Johnny Hansen, in front of the doorpost of Jordbrek.*[3]

---

3    Photo by Larry Hansen, taken 2002, the year his father, Johnny took his three sons to Norway

*It is very possible that the first planting took place around the year when the main building was raised. Two rows of cherry trees were planted on the southern hillside starting from the southern end of the main building and continuing some distance further downwards. Two or three rows of apple trees were planted at an angle to the shed which had been raised by my father in 1876. These had to be removed when the barn was built. But one row at the rear was left. These were unusually favourably situated, for nourishment as well as for sun and light. I still think I can imagine the enormous trunks and branches of these trees. But Hans Skjeldrum decided that these trees were no longer worth cultivating, so he had them removed. The same destiny also struck the cherry trees of the hillside, south of the main building. So many wonderful memories of childhood were attached to these cherry trees. When they were brimming with juicy berries in the summer, climbing on top of these tall cherry trees and eating ourselves full of cherries was like a paradise to us boys. [There were five boys and one girl in this family, like the family I grew up with, I just realized.]*

*\*\*\**

*The Konnerud road was completed in 1885. It now runs on the flat piece of land below the Jordbrek farms. The traffic around the Jordbrek farms came to an end. One would presume there was an interest to plant again, so that the orchard could be expanded further towards Skjeldrum. In fact there was a younger orchard there, consisting of apple trees, pear trees and berry bushes.*

*I must also tell about a novelty in the garden we were often made aware of. It was a pear tree, planted into the trunk of a rowan tree. Down towards the ground, it was a thin rowan trunk about 20 to 30 cm long, then there followed a considerably thicker pear trunk from this point upwards. We often looked with astonishment upon this novelty. It tended to burst into flowers at spring, but I can't remember we ever saw fruit on this pear tree. Therefore, it was probably chopped off and thrown on the flames a long time ago.*

*The entire garden of southern Jordbrek was surrounded by a hedge of common hawthorn. But, as it had not been taken care of by cutting, the thicket was tall and mighty and impenetrable, for people, as well as for animals.*

*\*\*\**

*As for the buildings of southern Jordbrek, I already mentioned the farmhouse and the shed. But I also want to mention the wash house, both because I can visualize it and because it was so old that it might as well be thought to have belonged to the first settlement at Southern Jordbrek, about 1755. It was situated at a 90° angle to the main building that was constructed later, so that the distance between garden fence and the washhouse was the width of a car.*

*There were two entrances on the north side of the washhouse. The main entrance was in the centre of the building and consisted of one double door leading into a small aisle. Inside, there were three doors. One straight ahead, led into what we called the wardroom of the hired hand* [where the hired help stayed]; *one to the right, led into the milk shed and the one to the left, into the wash house. This was big and spacious. At the right, was the open hearth, with a baking stove and a brewery pan for boiling mash* [a step in preparation for brewing beer]*.*

*Furthest to the right, was that door that led into what we called, the washhouse chamber. It was small. At the left of the open hearth was a double door as well, but this was cut horizontally, so that it left an upper door and a lower door. The intent of this arrangement must have been to let in light and air, while at the same time keeping the domestic animals out.*

*Further, in the washhouse there was a small square window at the left. The washhouse was heavily used during summer. As soon as we started the spring work, we moved out into the washhouse, and stayed the entire summer until the last of the crops of the year had been put under roof. To the right of the washing house, there was a workshop that was used a lot. The washhouse was torn down about 1910.*

*The new owner of Southern Jordbraek, Hans Skjeldrum, has restored the main building in a sensible and reverent fashion.*

***

*Of the many memories from my childhood that are associated with the place I grew up, there is one memory I would like to recount. I spent much of my childhood sick in bed. I remember one winter morning, I slept, in what we called the small wardroom, in the southern part of the main building. I was awakened by a strange sound- 'plonk, plonk' from the barn floor and this sound continued. When at last, my mother came into me with breakfast, I questioned what that sound was.*

*The explanation of the mystery: There was a man threshing rye at the barn. He was using an arrangement that was called a 'sliu' (probably connected to the word 'sla' which means to beat). It consisted of a pole of solid ash, about two meters long and four to five centimeters thick. At the lower end of this pole, a shorter pole as thick, was attached about 1 meter long. This was solidly attached to the longer pole by a strop of leather, so that it was able to slide around in a track lowest down on the long pole. This was an efficient implement to beat the grain kernels free from the ear of grain.* This would be like a flail.

*But already, during my childhood, more modern equipment, such as horse drawn grain harvesting machines were coming into use.*

***

I thought it was important to include the specific family history recorded in the Skoger book. This record from page 63 of this book lists one of several sets of my great- grandparents and their family members. The letters in bold are my direct ancestors.

Remember these are John's brother's words. *Here at this farm, dwelt my father and mother. My father grew up here at the farm, where the family has resided since the 17th century. My mother was from Gundersrod (another farm in the district of Skoger).*

My grandfather's grandparents [my great grandparents] were:

*Hans Brodersen Jordbrek and Anne Marie Andersdatter:*
*Maren Johanne–b.1843,*
*Ingeborg Andrine–b.1844,*
*Helle Birgitte–b. 1846,*
***Bernhard**–b. October 21, 1849, d. 1894.*

*\*\*\**

John's mother (my great grandmother) ***Elise Andrea Gundersrod**, was born April 17, 1850 to parents **John Tollefsen Gundersrod and Karen Kristine Eriksdatter**.* John and Karen would be my great-great grandparents. This record is from page 16 and 17 of the 'Blue Book.'

Elise's siblings (John's aunts and uncles from his mother's family) were:

*Talette Amalie–b. 1840,*
*Hartvig Edvard–b. 1846, Immigrated to Michigan, USA.*
*[J.B.Hansen's first home in America.]*
*Julius Christian–b.1848, took over the father's farm,*
***Elise Andrea**–b. 1850, Came to Jordbrek, d March 7, 1931,*
*Hans Martinius–b. 1852,*
*Gabrielle Christiane–b. 1854,*
*Emilie Randine–b. 1855,*
*Kristine Theodora–b. 1857,*
*Hanna Marie–b. 1858.*

JB Hansen's grandparents and aunts and uncles on his mother's side.

Standing l-r- Hans, Julius and **Elise Andrea** *(JB's mother)*, adults seated- Talette, **John Tollevsen, Karen Kristine Eriksdatter Tollevsen** *(was a widow with two children when she married)*, Hartvig, front l-r- Gabrielle, Emilie, Kristine, Hanna.

Elise Andrea Jordbrek

Bernhard Jordbrek

Elise and Bernhard are my great grandparents.

*The children of **Bernhard and Elise** were:*

Hans-b. January 15, 1874- d. December 22, 1944,
Anna-b. December 27, 1875- d. September 25, 1915,

**John**-b. April 7, 1877- d. January 27, 1965,
Karen Kristine-b. August 29, 1880- d. May 4, 1884,
Adolf-b. July 19, 1884- d. August 16, 1884,
Adolf-b. July 12, 1886-d. October 13, 1955,
Karl- b. March 8, 1889. Died as a child,
Broder-b. February 2, 1891-d. July 16, 1967.

l-r Broder, Adolph, their mother, Elise, John, Hans
during John's Norway visit 1925

\*\*\*

When I visited Norway in 1966, I went to Kongsberg to visit my second cousin, Bjorg, her parents, Berger & Astrid Severinsen, and her fiancée, Hermod Monsen. (They were married later that fall.) Bjorg's grandmother was my grandfather's only sister, Anna, and when she died in early 1900s, the fifth child born in the JB Hansen family was named Anna, in her honour. I was so surprised how much Astrid did remind me of my own Aunt Anna, in both her manner and appearance, and that her brothers I met, looked so much like my uncles.

After my aunt Myrtle and Evelyn's 1956 family visit to Norway, Bjorg and I, who were the same age, were encouraged to become pen pals. That relationship continues to this day. A cousin reunion was held at the Oungre Regional Park (close to the original homestead) on June 28, 2019 when Bjorg and Hermod and their family visited their Saskatchewan relatives as part of a North American tour.

*** 

My time in Oslo, Norway in 1966 was short. After scheduling the overnight ferry to Denmark, I had just enough time for a brief introduction to JB's brother, Broder. Though my pay phone calls to his daughter-in-law, Ruthie, kept getting cut off after one minute, I finally understood enough to take the correct bus and know the exact stop where I should get off.

There was no doubt the man waiting for me at the bus stop was grandpa's brother. As we walked back to their house, I felt awkward for I couldn't communicate much until some of the younger generation helped to translate. And it was a pleasure to meet his lovely wife, Astrid.

Now as I recreated and visualized JB's emigration journey, I realized I missed seeing Jordbrek. It is on my bucket list!

***

CHAPTER 2

# NORWAY TO AMERICA

*John from Emigrant to Immigrant*
*1903-1908*

John went to school in his home district. (This is the school he visited on his 1925 trip back to Norway, where Broder, his brother, was now the teacher.) John went on to attend Fosnes School of Agriculture in Lillehammer from 1896-1898, as had several others in his family. He worked for a while as an agriculturist at several farms.

John recorded in his memoir- *Life experiences from working on two farms and then as assistant, at Gusdal and later at Haugesund was a lot of help. When an unmarried man, at the last place I worked, wanted to go to America, we agreed, that I* (John) *should rent the farm. I gave notice that I'd quit my job. After I gave my notice, he changed his mind and was married. All at once, I decided to go to America.*

Now, in a bit of a dilemma, John wrote to his Uncle Hartvig Johnson who had emigrated and now lived in New Era, Michigan. He reported, *I got a lovely letter from [his wife] Aunt Jacobine,* [4] *and two months after losing the*

---

4   Chapter 12 in my book, **The Princess Doll's Scrapbook**, covered John's emigration story with his translated travel diary, beginning with boarding the emigrant ship leaving Oslo up till he had been in America for several years.

*farm (in Norway), I left Oslo at 10:00 am, May 22, 1903, arriving in New York on June 6 at 4:00 pm, having sailed 15 ½ days.*

*When we landed, we all had to go by a doctor who stared at us and then, we were all herded into a large room with men, women, and children.* (This was to check for any communicable diseases. If any was found, the immigrant would be detained in a hospital facility until they recovered and were able to travel.) *Here we were locked into different rooms until they took us, in groups, to different trains.*

***

*At New Era, I was met by two cousins, Ernest, and Edgar Johnson, with the large wagon they used in their milk business, pulled by two horses. When they drove carelessly, I shouted, "Pro," once or twice. The small boys had a good laugh because here in America, they say, "Whoa!"* [Sounds like his young cousins had a bit of fun at John's expense.]

*I had a bad cold when I arrived, but with my aunt's loving care, I was soon better. When she suggested I change my name from John Jordbrek to John B Hansen, I wrote, John B Hansen Jordbrek, for quite a while. Finally, I thought it was better to drop the Jordbrek.* (The "B" was for his father, Bernhard. Under the Norwegian patronymic naming system, his last name would have been Bernhardsen. Around this period, that traditional naming system was being discontinued. But John's grandfather was Hans, so his father's last name already was Hansen.)

*After I came to Estevan, I would have liked the name, Jordbrek, for there were so many Hansen families, the mail often got mixed up.* And Jordbrek would have been a good name for a farmer.

*I worked for Victor Munson at New Era at $20 a month, where I cultivated the tall corn with a big red horse without reins in the tall corn. I said "Gee" for right and "Ha" for left.* [That was a well-trained horse.]

***

*I went to Slocum, Michigan, where I worked at a sawmill. My first job was carrying lumber all day for $1.75 per day. Later, I got work loading a trolley with newly sawn lumber. I liked the work, but one day I slipped on ice and broke my left thumb, so* [I] *went to a doctor. It wasn't set right, so it has gotten stiff lately.* [I think he is referring to how it feels now, in his senior years, for don't those old injuries seem to come back to haunt us?]

*After I broke my thumb, I got work calculating how much lumber had been sawed each night, to hand in to the office in the morning. Arithmetic had always been easy for me. I enjoyed that part very much, but not the night work.*

\*\*\*

*In the summer of 1904, I decided to go to Martin Brekke's at Lake Park, Minnesota. I got work at one of his neighbours in threshing. I stood there throwing bundles into the threshing machine from early morning till after dark, for $1.50 a day, and I had to find my own food and bed in rainy weather. Had it not been for Martin Brekke, from the neighbouring farm back in Norway, who gave me both food and bed, all I earned would have gone for my upkeep. And this was the best time of the year for work. When I complained, Martin said, "You must go to school, John."*

\*\*\*

*When threshing was over, Martin's two sons, Henry and George, and I went to Aakers Business College in Fargo, North Dakota. We did light housekeeping at John Brekke's, who was Martin's brother, an old bachelor, who lived on the east side of Moorhead. He was kind and all went well. It was a busy time, as I took English at night, too, and so we had to make three trips, back and forth, as we also went home at noon.*

*I saw a fellow I knew I'd seen before and finally remembered, he was John Harilstad, who I'd met at his uncles. When convenient, we had a good chat about home in Norway.*

Isn't it special to meet someone from 'back home' when you are in an unfamiliar place?

## ELAINE MELBY AYRE

\*\*\*

*In the spring of 1905, I began to work for Huntoon, a banker in Moorhead, MN* [just across from Fargo, ND] *who had quite a large farm, south of Concordia College. He had cows, purebred horses, and he bought wild horses that we tamed alongside the purebred horses. I really liked breaking wild horses.*

*My brother, Adolph, left Norway April 6, 1905, and arrived in Moorhead, April 22. He worked for Johnson, a Swedish farmer, eight or nine days and then got work with me at Huntoon's, for $140 for 6 ½ months.*

*In the middle of the summer, we bought a small dairy and hauled milk to Fargo, selling it at 16 quarts for $1.00. Now* (January 1957) *we get five quarts for a $1.00.* [In 2017, a liter, a cup less that an Imperial quart, sells for $2.50.] *We were well paid for our work even if we had to buy straw for bedding.*

It appeared from **The Skoger Book**, that back in Jordbrek, my grandfather, John's home in Norway, dairy farming was a business they were familiar with.

\*\*\*

John continued, *"The next summer* (1906) *we sold the milk business– the cow barn and cows and took our four old horses and wagons and got a railroad car with a Mr. Munson and went to Flaxton, North Dakota.* (Flaxton was a community about 80 miles northwest of Minot.) *We hauled what we had by wagon from there to Fortuna.*

Fortuna, located further west along US Highway 52, was the closest border crossing point to the Hansen Homestead, across from the Canadian site at Oungre on Saskatchewan highway # 35, though the area history book indicated there had earlier been a crossing point, south of Ratcliffe.)

*We spotted some land and drove to Williston with our old horses to file on the land and bring back lumber. Munson and my brother, Adolph, got their papers on the land they had spotted but what I had picked was already taken, so we brought back lumber for my brother's little 10x12 house.* (On one of the last visits I had with my uncle Lloyd, he told me that, but for John losing out on this piece of land at Fortuna, our family might have been American.)

*We had a small tent we slept in, as we built the house. After we built the house and put-up hay, we took the plow and made two or three furrows about thirty feet apart and made a fire to burn up the dry grass between the furrows for a fire guard. It gave us a trying time when the fire jumped the furrow and the four of us worked for our life and got it out. We were so tired; we fell down and slept there for hours.*

***

*Next, we all drove east for threshing. Being a little early for threshing, we lay over at Portal for a few days. One night, it rained and blew so hard, we had to dig in our heels to hold the tent over our heads until morning. We all got work on a threshing rig, north of Portal, and when done, headed back to Ambrose, where I bought a building site and built my first 10x12 foot house and a barn for the horses.*

*I got work to build a cellar under a large store, and later, to haul lumber for many homestead shacks for new settlers. One time, when I was coming back with an empty wagon, south-west of Ambrose, a prairie fire came towards me. I stopped, took a match and set fire to the long grass, and then drove in on this burned patch till the fire was by me. (That's what you'd call 'fighting fire with fire')*

*Later, as I and the fellow I hauled lumber for, each drove a load going south of Fortuna, rain turned into a real North Dakota snowstorm. We decided to unhitch and ride our horses back. With our stiffly frozen clothes, it was difficult to get on the horses. It was more than funny.*

*We knew the horses could find the way, and 'Praise God' before long, we were at a sod house and barn belonging to a fellow we had threshed for. The bachelor got a two-day visit with us.*

*We drove to one of his neighbours to see about hauling coal. In a low area, one of the horses jumped sideways, and as it did, the horse stepped on the strap holding the neck yoke. It broke and the pole between the horses hit the ground.*

*We jumped, and the next minute, the buggy went flying through the air. One horse died on the spot, when the sharp neck yoke hit its chest. We never got much coal hauled that time.*

\*\*\*

*Since it looked like winter was here for good, I decided to go to Moorhead and see if it was possible to get what we had coming on the cows and barn. We were fortunate to get out on the last train to leave Ambrose* [that season]. *Because of all the wind and snow, the new train that had come to Ambrose in 1906 was useless until the spring of 1907. Many suffered hardships that bad winter. Two ladies had moved out, thinking they had all their supplies. The storms started, and having no matches and consequently, no heat, they froze to death.*

*I couldn't find the man I'd sold the cows to, so* [I] *was out that money and very disappointed.*

*I visited Martin Brekke and family on the return trip---also Christ Holm at Barton, North Dakota who also was from* [my home area] *Skoger, Norway. He was Martin's sister's son. He loaned me some money, which came to good use when our train got stuck in the snow at various places. I got to Flaxton and got a job clearing snow from the railroad for the blacksmith.*

*I took off one day with a man with two horses pulling a sleigh. We hadn't gone far, before we were in a terrible snowstorm. When he stopped the horses, he got out and started to go in circles, I hollered, "What is wrong with you? Come into the sleigh."*

*He did and "Praise God" it wasn't long till the horses got us into a small town. The next day, we got to Ambrose, where my brother was doing fine, but three of the four horses were dead in the barn.* [I didn't know if this was because it was such an extreme winter as was mentioned before or if it was due to them being old horses or more possibly it was some horse disease.]

Page 141-143 of my previous book, **The Princess Doll's Scrapbook,** provided further evidence of that bad snowy winter of 1906-07.

# HOMESTEADING IN THE LAST BEST WEST

*** 

*Now, all we had was our own energy to work with, so I began to haul hay for the livery barn. (A livery was a place in town where farmers could rest and feed their horses). I got [paid] $1.50 a day- better than nothing in the winter.*

*Ragnvold Hagen, a new settler had gone into a cement business, but when he couldn't make the payments, I bought him out and I had a good paying business- even hiring extra help.*

*Later, I got work in a lumber yard, making racks to place lumber on. I soon became second man and was helping to sell and load lumber. I quit this job and got the same work at a newly built larger lumber yard, Kulaas Lumber. I became the agent and moved my belongings from my shack to the fine sleeping room beside the office. It was a help to have worked at a sawmill in Michigan, as well as to have had the course at Aakers Business College.*

*Since I like working with numbers, I enjoyed figuring out how much lumber of each kind the buyer needed for his building. Since, I earned a certain percentage [of sales] –days became very long.* It seems JB maximized earnings by working as much as he could.

*Since I'd had hard times, I decided to be [as] helpful as possible to the new settlers. I let them bring in their horse blankets and sleep on the floor of my office. Some nights, five or six men were asleep on the floor when I finished doing my books at midnight. I had to be up again at 5 am, loading two or three loads at once for the settlers, as they had a long way to go home.*

*I used to put the money in my jacket. One day I was short $20, so had lost it from my pocket.*

Losing his hard-earned cash was a problem worth recording, but now *as he goes back and forth from his work, John's eye caught sight of a lovely Norske girl in the neighbour's window.* He is about to gain the love of his life.

***

CHAPTER 3

# JOHN FINDS A HELPMATE

*Ambrose, ND 1906-1909*
*Courtship, marriage and homesteading*

John's next recorded impression of that lovely Norske girl, destined to be his wife and my grandmother, happened this way: *When I went to the church service, held in the railroad station, I got the chance to see Mrs. Gronvold and her daughter, Dina. I was really impressed by how well they sang.*

*I soon started visiting Dina and took her out on various occasions. In the summer of 1908, Dina and I went by buggy all the way to Plentywood, to visit her three brothers, who had homesteaded there.*

\*\*\*

*That fall, Even Ulledalen took me to Estevan to file on land – a homestead and a pre-emption.* [The date was September 16, 1908.][5]

---

5      There is a picture of that group of prospective homesteaders lining up outside that office to keep their place in line on page 148 of **The Princess Doll's Scrapbook.**

> ## Pre-emption
>
> It sounded to me like John's friend, Even was aware of recent changes in the homesteading act they could use to their advantage.
>
> A pre-emption was a process adopted from the United States in 1874, allowing a settler who had entered a homestead to obtain an "interim entry" on another quarter-section located adjacent to his homestead. After he received his Letters Patent (or ownership) for his homestead he could then purchase the additional 'pre-emption' land at government prices. Pre-emptions were discontinued in 1890, reintroduced in 1908, and repealed in 1918.[6]

*I filed on land, by proxy, for my brother, Adolf. To prepare to go on the land in the spring, I purchased some used machinery and two unbroken bronco mares. So, some* [of my] *trips* [going back and forth to get things set up] *were speedy. Johan Gronvold and his son, Olaf, had also taken up land in Canada, right next to ours, so Henry and Jens took out lumber (from Plentywood) for my* 12x16 shack.[7] Though John didn't say, my uncle Bernhard had told my cousin, Larry that the homestead shack had been built in the fall of 1908.

*\*\*\**

> ## "Postals"
>
> When the family cleared out my Aunt Evelyn's things after her death in 2012, I asked for my grandmother, Dina's scrapbook collection of early 1900 postcards. I remembered studying them

---

6     https://familysearch.org/wiki/en/Canada_Land_and_Property

7     There is more of the background of Marie and Johan Gronvold recorded on page 150 and 152 of **The Princess Dolls Scrapbook.**

when I lived at my grandmother's place from 1966-68. These little notes provided interesting personal news from the period after the Gronvold family first moved from Fertile, MN (where Dina was born and raised) to Ambrose, ND and on into their first years of homesteading. The ones that add to this story will be shared. With limited space, the writing is tiny and abbreviated, but not like today's instantly received text messages. These postcards may have taken several days or a week or more to reach their destination. Many of these "postals," as they called them, were from Dina's new sister-in-law, Mathilda Morvig, who was now married to her oldest brother, Jens Gronvold. Dina's older brothers, Jens, Henry and Adolph had homesteaded near Plentywood, in the northeast corner of Montana. And since the Morvig family had been close neighbours to the Gronvold's near Fertile, MN, they had all known each other their whole life. Jens and Mathilda were married back in Fertile, MN on July 1, 1908, nine months ahead of John and Dina's March 25, 1909, wedding.

On August 18, 1908, on completing the return trip to their Plentywood, MT homestead after their wedding back in Fertile, MN, my great-aunt Mathilda writes to her sister-in-law, my grandmother, Dina

**Sister-in-law: We got there all right Friday evening but late when we left Melby's. Adolph took the cows and went on first. He got here before dark. The black cow gave up, so we let her go loose after. Garden and flowers are nice and the oats.**

At this time, the halfway house between Ambrose and Plentywood was Rolson where my paternal grandparents, the Dahl Melby Family operated a country post office, and that was where Jens and Mathilda stayed. It was common that people who lived along the usually travelled routes would offer overnight hospitality. There weren't hotels and motels along the way.

When Jens and Mathilda visited at our Melby place in the late fall in the early 1950s, they joked about this event with my father. For at a previous time, before they were married, they had passed through, and Mrs. Melby had put them up in separate beds. She was about to do that this time, and now Mrs. Melby was embarrassed for she hadn't realized they were now married!

The Dahl Melby family homesteaded in Saskatchewan around the same time as the Hansens did

***

Next, my great-uncle Henry, Dina's brother, informs us about some previously unknown facts about the situation their family faced in the days leading up to John and Dina's wedding. Henry is back visiting in the Fertile, Minnesota area where the family moved from several years before. He writes:

**March 8, 1909, Dear sister Dina, I received your letter and postal a while ago for which I thank you. I had a letter from Hansen yesterday, he said you were quarantined out there …. we've been expecting a long letter from you for a while, but I see you can't mail any letters while you are quarantined. I also had a letter from Jens yesterday. Your brother, Henry**

Then a postcard from a friend, possibly this was one of Dina's cousins back in Fertile, Minnesota, tells us exactly what was happening:

**March 29. 1909, Dear friend, I would have written to you before but it seems like my time is taken up with something else all the time and so you have had diphtheria. How is your ma? Remember me to her. I would like very much to visit you but don't know when I can get there now. Someone tells me you work in the post office in Ambrose last fall, did you? Your friend, Mrs A. Goodale**

***

John continued: *We were married on March 24, 1909, in Ambrose. Dina's brother, Jens Gronvold and his wife, Mathilda, were our attendants. I do remember well the Pastor's words, in Norwegian, of course: "As for me and my house, we will serve the Lord."*

A small newspaper clipping Dina saved, announced their marriage, stating simply, "The wedding was a quiet affair", not surprising, considering the family had been sick.

*John & Dina in their Travelling Outfits, looking the part of a power couple.*

Their wedding portrait was featured on Page 14 of The Princess Doll's Scrapbook. Dina's wedding dress was the inspiration for that doll's dress.

And about a month after John and Dina's wedding Mathilda wrote about the progress they were making at their homestead, **4/27/09, Your letter at hand, was glad to hear from you, thanks for the same. How is mama? Hope she is all right. We are all well and are done seeding and started breaking, they have soon broken 5 acres. Henry will soon go to Ambrose, as soon as the trees** (come in.) [This could either have been trees planted for a shelterbelt around the homestead or perhaps it was referring to The Timber Culture Act of 1873.[8] This was an addition to the Homestead Act in the United States and granted land to a claimant who was required to plant trees- and that tract could be added to an existing homestead.] **I forgot to tell you in the last letter that Ole Morvig came here the same evening we got here, and he stayed till the next day. Mathilda**

And then, there was this undated card from before they left Ambrose, with the salutation "sister and brother-in-law." I thought it was likely in response to a gift they'd been given for being in John and Dina's wedding party. I imagined it had been sent in the mail along with flower seeds from the Gronvold store (for where else would flowers be available ?) for it says: **Yesterday I got my flowers and when I opened the box I found something wrapped up in a napkin. I tell you, I got surprised. You should not have done that cause that was entirely too much. We thank you so very much for it and all the pictures, it was really nice, thanks. With lots of love to all, Mr. & Mrs. J.C. Gronvold**

From now on, letters were addressed to Mrs. J.B. Hansen, Byrne Post Office, Sask. Canada. This local country post office was operated by Philip Byrne, who along with his wife, Catherine, homesteaded in Souris Valley Municipality in 1906, about three to four miles northeast of the Hansen homestead, 20 years before the villages of Hoffer or Oungre were established. Prior to that, the mail point had been Hamar, a district, south of Bromhead. The post office might have been named Byrne, except for the fact that the Jewish Colonization Association donated a sum of money to build a community hall if the village be named after Louis Oungre, who was then current leader of that organization.

***

Now, we begin to get a picture of homesteading realities as John shares the beginnings of his family and farming operation.

***

---

8     https://en.wikipedia.org/wiki/Timber_Culture_Act

CHAPTER 4

# HOMESTEAD HONEYMOON

*NE 23-2-15 W2*
*R.M. Souris Valley #7, Saskatchewan*
*Ten miles north of the US Border*
*Western edge of North Dakota*

The beginning of my grandparent's (John and Dina Hansen's) life together on their Saskatchewan homestead was June 1909 according to the family history recorded in 'The Skoger Book.'

\*\*\*

*We were soon off to our homestead in Canada with our earthly belongings in a wagon drawn by an ox and a blind mare. We arrived safely at our honeymoon home.*

*I bought an untamed ox. I put Buff in front, and the untamed ox had to follow him as I was breaking the land. After making a round or two, I let Buff steer it all, and I followed behind, digging up stones and throwing them aside. After I'd done this from 4:00–10:00, I let the oxen loose to eat grass and rest.*

I wasn't sure if he got up first thing in the morning to do this or if he was referring to the late afternoon to evening hours, but in what he wrote later, it appears he was an early riser.

*Our first crop just barely covered the threshing expenses.*

***

John's memoir begins to explain their changing circumstances as the first family members are born, beginning with my mother, Edith, and a year and a half later, her sister, Myrtle.

*Edith was born March 3, 1910, and Myrtle was born October 31, 1911, so there got to be both babies- crying and singing and more work.*

In one of many of my mother Edith's little history bites, she explained how their family managed things when the crop barely covered their expenses. "Father spent some of the off-farming season, working at Kulaas Lumberyard in Ambrose. But since [my] mother's parents and brother, Olaf owned a country store on the cornering quarter, mother could easily walk over there when she got lonely, and the work was done."

And it was a fact, many homesteaders required paid work outside the farm so they would have enough resources to get started in their farming operations. John was fortunate for his experience and earned pay working in the lumberyard earlier was a help to them as they were getting started.

***

As John continues, we can see that they are becoming familiar with the wider area. *Late in 1910, Mike Kleven and I drove* [about 8 miles west] *to Dahl Melby's to bring home cattle that had been near there in pasture all summer. A prairie fire came very near their place, but Mike and I put it out, and saved their house.*

*My paternal grandparents,* the Dahl Melby homestead was about 8 miles straight north of the North Dakota/Montana state line. This will give you a better idea of Hansen's homestead location.

"A black countryside with multitudinous white rocks, that had to be picked before one furrow of ground could be broken,⁹" was how the Emil Torkelson family, close-by neighbours of my paternal grandparents, the Dahl Melby's, described their first view of their part of the southern Saskatchewan landscape, on their arrival in the spring of 1911. This story, as well as others I had read in the community history, emphasized what a threat fire was for homesteaders in those early years before the prairie was widely cultivated. In fact, even before building, their first task must be 'to plow a fireguard' as we have seen in what John wrote of his experience.

So, John had been part of that story, and now further adventures of that day continue. He recalls. *As we headed home, we saw thick smoke towards our home, where Dina was alone with Edith. Dina had taken her in the carriage and gone to her folks,* [then she had] *ridden horseback to get the men* [who must have been working out in the fields] *and they put out the fire near our home. God is so good.*

***

Correspondence with the Gronvolds continued, and based on the progress in their part of the world, we see the reason for Mathilda and Jens to be excited:

**/17/ 1909… Well, we are going to get the railroad now, they have already commenced grading down at Pete Masson on the Muddy [Creek] and have a large [construction] "camp".**

**Had a birthday party on Sunday, will tell you about it in the letter. Best wishes, Mathilda**

[The Big Muddy was a badlands area in Southern Saskatchewan, north of eastern Montana where Plentywood was located.]

---

9     *The Settlers of the Hills* Community history book for the Rural Municipality of Lake Alma 1976, from the Emil and Hulda Torkelson Story, Page 257

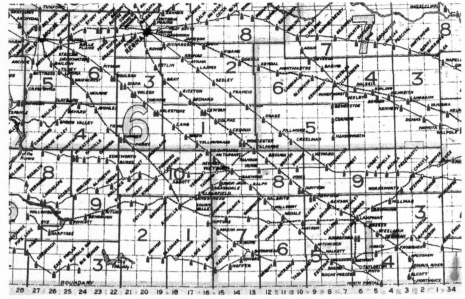

*bottom southeast section of the Saskatchewan Wheat Pool calendar[10] shows the rail line from the late 1920s*

## Municipality of Souris Valley No. 7

Meanwhile the Hansen's still had a long wait for the railroad to be close by. The first rail line in this area of Saskatchewan came into operation in 1913, between Estevan and Neptune with Bromhead and Tribune in the middle. Tribune was most likely where they did their grocery shopping. But the long anticipated Canadian Pacific branch line, from Bromhead west out to Minton, was not realized until 1926 and the villages – Outram, Torquay, Oungre, Hoffer, Ratcliffe, Beaubier, Lake

---

10      Adamson, Julia (2002), 1947-48 Saskatchewan Pool Country Elevator System MAP with calendar index, https://sites.rootsweb.com/~skwheat/Maps.html, Saskatchewan Wheat Pool Maps 60 Years 1924-1984, May 2002, retrieved October 4, 2019 Permission granted for reproduction from Viterra, the company that took over from the Saskatchewan Wheat Pool.

Alma, Blooming, Gladmar and Minton were the points along the rail line. Several of these hamlets no longer exist.

A rail line, from Goodwater to Colgate, shown on the 1915 map of the municipalities of Souris Valley and Lomond, is named Maryfield Extension[11] had been built earlier, for according to the Tribune town history, when that town site was surveyed before the rail was laid, building materials were hauled from Colgate. Tribune was ten and Colgate was twenty miles away from the Hansen homestead, but that was still an improvement over the forty-mile trip it had been to Ambrose, the place where they did their buying and selling in the first years.

A 1912 photo of Tribune showed a hotel, implement shop, blacksmith shop, real estate and insurance office, Lee Hing Café, printing office, store and barber shop and Sprague's store, which was operated at various times by my paternal grandfather, D.L. Melby. This would probably have been around 1915, for my father, Victor told us once as we drove past the large slough, south of the Tribune town site, that he remembered skating there, when he was about six years old.

When the rail-line finally came through to Hoffer, it was barely a three-mile trip to town from the homestead. Now that line has been taken out and farmers are back to hauling their grain 40 miles (74 km) to the inland terminals. (Mile was the unit of distance measurement used in those days. A kilometer is 0.621 miles.)

That 1915 map of the Souris Valley and Lomond Municipalities[12] was bordered with 29 advertisements for a

---

11     **The Saga of Souris Valley RM #7** a community history published by Souris Valley No. 7 History Club, Box 22, Oungre, Saskatchewan, Printed by Friesen Printers 1976, from Village of Tribune history. Page 667.

12     The full-size wall map of Souris Valley RM No 7, 1915, received from my brother, Palmer Melby, from when he farmed in Souris Valley RM No.7. Map was also on inside front cover of the community history, **Saga of Souris Valley RM #7**. The replica copy of this wall map was originally published by Geo. W Atkinson, of Ceylon, Sask., part of The Municipal Series.

range of businesses including The Colgate Garage and Livery, who specialized in oils, greases and Ford Auto accessories and that 'refueling was a specialty', various merchants of dry goods and groceries, general blacksmiths, a breeder of Holstein cattle and one of Clydesdale horses, and so on. Weyburn Security Bank advertised branch offices in Tribune and Colgate.

Sections of land within the patchwork of rural school districts also highlighted the natural features of the land, like coulees, sloughs, creeks, trails, and even the springs in the coulee and the hill southwest of our farmhouse where my brothers and I used to go tobogganing. It was part of section 17 and 20 on the place marked C.J. Anderson at the western edge of Lyndale district on this early map, the next district east of Dravland. By 1946 it became the Victor Melby farm, where I grew up. You can see these features better on the eastern edge of the larger version of the Dravland School District in Chapter 9.

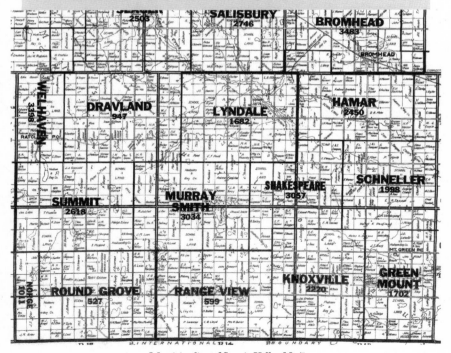

*Municipality of Souris Valley No. 7*

Mathilda's next note informs us that Dina's brothers had made a trip from Plentywood to visit John and Dina, to mark the first holiday season at their homestead. Plentywood was at least forty miles away and there seemed to be an ongoing query about Mama (Maria Gronvold, Dina's mother) who I thought likely had some health issues: **Jan 4, 1910, Sister in law, Your postal at hand. Thanks, was glad to hear from you and that mama was better and that H** (Henry) **and A** (Adolph) **came there all right. Have had a cold spell now and a blizzard, but today it is trying to snow. I suppose the boys have left by this time. We are well except we have a little cold. -?- has been here to visit us and now he came here New Year's Eve and visit again January 2 and we had company here New Year's day, it was these neighbours of ours M.C., A.O., O. L.** (I thought it was an inside joke) **and Mr. Helgaas and Miss Lutness!....** [We] **were to a Xmas Tree the second day of Christmas in Plentywood.** [Probably this was some type of Christmas program.] **Wishing you a Happy New Year. Greetings to mama and all"**

\*\*\*

**24/1910** [I] **should have answered your letter today, but have not any paper nor envelopes at hand, but will get some today. We are to have** [church] **services here at our place next Sunday, the 28th. Are you coming? Best wishes to you all, from us. Greet Ma and Pa from us.**

A 1910 calendar showed the date, Mathilda referred to, was August 28, 1910. John and Dina had just been there during July when that first photo of Edith, shown below, was taken. So how long did the mail take? It seemed that the anticipated time for a letter to arrive could not have been that long if they expected John and Dina to come at such a short notice.

\*\*\*

Telling about their travel in those times, John's memoir continues:

*We drove to Plentywood with two horses and a buggy to visit the Gronvolds.*

*When we got there, Henry was out courting, and I remember saying, "I wish Henry would come so we can see his 'sweetheart' before we leave."*

*Our wish was fulfilled, when he came in a little while with her, in a 'top buggy' with two horses. At the dinner table, we spoke of the Grimsruds, and I said, "That was a name from where I came in Norway," so Helen Flaten asked where I came from.*

*When I said, "Skoger," she sat up, and I said "Jordbrek."*

*She hit the table, so the dishes danced, and said, "You aren't really the son of Elise Jordbrek?"*

*My dad and hers were related and* [they] *had lived on a neighboring farm and had gone to school in our dining room, which had been used as a school for several years.*

Helen should be recognized as an especially gutsy lady for she homesteaded as a single woman in that area on the piece of land which became Henry and Helen Gronvold's home place. Helen and Henry were married at Antelope in the Valley district near Plentywood on Feb 5, 1913.[13]

*We visited Gronvold's, near Plentywood regularly over the years. These studio photos following are the evidence.*

\*\*\*

---

13   E-message from Larry Hansen Feb15, 2021

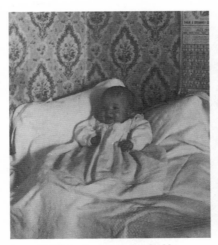 

*Edith is about 4-5 month old, as the calendar shows July 1910.*   *Edith, Myrtle, Bernhard and Clarence 1915*

And about a year later, in the postcard "Harvest time on a Dryland farm" shown below, Dina's brother (my great uncle) Henry sounds more than a little annoyed at his brother, Adolph.

**Jan 10 1911, Sister Dina received your postal……. getting along all right, have plenty of work to do, have not been to Plentywood for one and a half weeks now. J. and M. (Jens and Mathilda) got very busy up there when they don't come back. Heard from Adolph some time ago, he wanted money -he wanted $30 to come back with. The fare must be very high, or I don't know what. Brother Henry.**

*Collage of three postcards. Top: the breaking plough, the man at the back with the beard is Johan Gronvold. Middle: Henry's postcard illustrates cutting grain and stoking. The cancellation stamp on the other side is Plentywood, Jan 10, 1911.*
*Bottom: the undated postcard, whose message appears in the middle of the next page.*

Then a note from Mathilda informs us of some long-forgotten family history, and gives us clues as to why Adolph was delayed on a visit back to their old home area around Fertile, Minnesota,

**02/19/11 Hello Dina, received your card and Valentine, which I thank you for. S'pose you have received my letter by this time. Yes, should say it was nice to get home again. I suppose I told you how many postcards and letters that was here when I came home- 25 postcards and letters yesterday. I received the letters Olaf forwarded and the folks at home has been sick--- Gunner has had the lung fever but is better. Clara's little girls have the measles and Hannah has been sick, had 103 fever and the rest has also been sick. Adolph has not come home yet, I don't say.**

(Mathilda's sister, **Ida,** who lived back at Fertile, Minnesota) **said he was going to leave this week and he was to have his girly along.** (Adolph must have had a prospective love interest at this time, but he was a bachelor for his whole life) **Best wishes to you and all from us. Have sold three cows, butchered two steers and sold part of the beef. Sister-in-law Mathilda.**

And then there was an undated postal from Mathilda, with her married name, Mrs. J.C. Gronvold, inscribed in gold ink on the front of the card but it was probably sent in an envelope as there is no postage or cancellation stamp, and tells they had visited John and Dina in Canada. I thought that must have been sometime in the winter of 1910 before my mother, Edith's March 3 birth:

**Hello! How are you folks getting along? Can say we are all well and hope you are the same. Can say we have some snow now but it is getting milder, so I think it will go soon again. But Sunday we had a real blizzard from the east, it is the worst one we have had since we came back from Canada.**

**Best wishes from Mathilda. How is Mama?**

***

Now, John relates more of day-to-day life on the homestead from those early years. *Once in a snowstorm,* [during the time] *when my brother, Adolf, and Mike Kleven used our kitchen for sleeping, they went out to carry water to the cattle, east of the barn. It seemed to be taking them too long, so I ran to the barn and found they had just gotten there.*

A prairie snowstorm could be very disorienting, a person could wander in circles and eventually freeze to death. In addition, the way he refers to where they had to go in the earliest years, the first barn was east of the house, where their first water source was located. Putting a rope from house to barn was a way to make it safer to go back and forth to care for the animals in the event of blizzard conditions.

Besides being careful in their own yard, I found it surprising that John and Dina and her brothers were making these long trips back and forth between Plentywood and their Saskatchewan homestead in the winter.

***

Now John details some of the matters related to breaking the land and getting water.

*I bought several oxen for breaking but it was so slow. I sold the oxen to Elving, who broke the land for me. When* [my] *brother-in-law, Olaf Gronvold, bought a large breaking machine, he soon had the rest of* [our Section] *23 turned over."* And it was now ready for seeding. He was getting ahead of the game in fulfilling the requirements of the Homestead Act. You saw a picture of a breaking plow in the postcard collage shown previously. It's quite possible the people in that picture are members of the Gronvold family and some of the Hansens' neighbours.

> ## Proving up
>
> To prove up the homestead[14], various obligations had to be met before the patent, or ownership of the land could be transferred from the Crown to the homesteader.
>
> The first requirement was building a habitable house and living there at least three months, then break at least 5 acres the first year. The second year, they must crop at least the first 5 acres, and break and prepare to crop no less than 10 more acres. Within the third year, the settler must crop the acres broken in the first two years, and prepare for crop, no less than 15 more acres. They also had to prepare to become a British subject [become a Canadian Citizen].

"Water used to be a problem. Wells were dug in every slough nearby. When the wells gave out, the cattle had to be taken several miles for water. Mother used to be good on horseback in those days." [15] These additional details came from another of my mother, Ediths' notes.

And John wrote, *"At first, I dug a well in a low place east of the shack for water but when I got more horses and cattle, I had to haul water in a big tank for two miles.* The answer as to where this was located came from my cousin, Larry Hansen. He had lived across from the Keives place, a mile or so northwest of the Hansen Homestead. He said that George Keives talked a lot about the fact that in the early days, people came for water there, near his place where there was a flowing spring.

*Later, I had a deep well drilled by a machine* [west of the house]. *They put down 2-inch iron pipes with 1-inch pipes inside, through which water is pumped by the windmill. This well cost $800, but after a short time we had no water, as fine sand filled the well. When the well driller returned to put a screen*

---

14   https://saskarchives.com/collections/land-records/history-and-background-administration-land-Saskatchewan/homesteading

15   The Saga of Souris Valley RM No 7 page 107

*in the well, I refused to pay, with the outcome that I had to pay $300 expenses, besides the cost of the well. There I was, with no water. So, I decided to buy a well drilling machine and hired an experienced man who got the well working. Now we've had good soft water for over 40 years. We drilled several wells around and made good, even if some wells weren't paid for.*

\*\*\*

The family increases and so does the work. Everyday life was not easy. It makes a fascinating study to compare what their daily lives were like compared with our lives today. Now, we'll look at how some of the basic necessities of life were met in those early years.

Chapter 5

# THE NITTY GRITTY OF DAILY LIFE

*Homestead basics 1909–1919*
*NE 23-2-15 W2*

I estimated that by the time John and Dina reached their 12'x16' honeymoon shack in 1909, John had already built several 10'x10' or 10'x12' homestead shacks, including one for his brother, Adolf and the other, his own accommodation, when he lived back in Ambrose, ND. This was in addition to all of the building packages he put together for homesteaders, who came by way of Kulaas Lumber in Ambrose, ND where he worked before homesteading on section 23 of the Rural Municipality of Souris Valley #7 in southeastern Saskatchewan.

Such a small space (10x10), equivalent to the size of a small bedroom in many houses of today, allowed for only the basics: a stove for heat and cooking, a table to eat at and use as a preparation area, a chair or two and a bed, some hooks, most likely nails to hang coats and clothes on. And of course, there was Dina's trunk where special possessions like her wedding dress, her travelling outfit and her treasured porcelain doll were hidden away. Their 12x16 homestead shack probably was two rooms.

A **Winnipeg couch** was versatile enough to double as a seating area and a bed. We had one of those in our house in the fifties when I was growing up. It started out as a couch on the side of the dining room. Occasionally, it became a single bed when we had guests and with a slipcover over the frame, it was a reasonable facsimile for furniture.

Later, when my brothers had to share one bedroom, it became an extra bed for two of them. The folded-down wings on either side of the couch were pulled up, legs folded down from the frame and the mattress was unfolded to make a double bed. It was barely a couple inches thick. Undoubtedly something like that was part of the first furniture in my grandparent's home for where else did John's brother and Mike Kleven sleep in John and Dina's kitchen, as mentioned in the storm incident in the previous chapter?

The table might have been quite rough, but it was always covered up with an oilcloth, (no plastic yet). That cover could be wiped off to keep it clean. When it wore out, it was replaced on a shopping visit. I remembered general stores in my time still carried rolls of it. Several different oilcloths covered our kitchen table when I grew up.

The corner, by the door, was a likely location for a washstand, with an enamel basin for washing and beside it, a water pail or stoneware crock to hold the water supply. Hanging on a nail on the wall at the side or resting on the crock cover was a dipper, used as a common cup whenever you wanted a drink of water. Beside the stand was the slop pail where used water was discarded after use. During growing season, wastewater did double duty- it was used to water the garden or the flowers. Another pail collected vegetable peelings and other food wastes- it was extra food for the pigs or chickens.

## How to Build a Fire

One place I lived with my family, 25 years ago, west of Water Valley, AB came with a wood stove. One winter, when the element on the electric stove gave out, using the wood stove for cooking our meals became an adventure for my family.

"How to Build a Fire" from **The Fanny Farmer 1896 Cookbook**, [16] explained the skills necessary to keep a fire going in a cook stove. Without thermostats or knobs to turn, attention to these fire-building skills kept the house warm and provided heat for cooking meals. How do you think you would manage?

*Before starting to build a fire, free the grate from ashes. To do this, put on metal (covers). Close front and back dampers and open damper; turn grate and ashes will fall into the ash receiver. If these rules are not followed, ashes will fly over the room.* And that would be a mess!

*Turn grate back into place, remove the covers over firebox. Cover paper, (that you have twisted in the centre and left loose at the ends) with small sticks or pieces of pinewood, being sure that the wood reaches the ends of the firebox, and so arranged that it will admit air. Over pine wood, arrange hardwood; then sprinkle with two shovels full of coal. Put on covers, open the closed dampers, strike a match, –sufficient friction is formed to burn the phosphorus, this in turn lights the sulphur and the sulphur the wood, –then apply the lighted match under the grate and you have a fire.*

*Now blacken the stove.* [Stove black is a polish used on wood-burning stoves, meant to restore the original shine by removing burnt marks and residue.] *Begin at the front of range and work towards the back; as the iron heats, a good polish may be obtained. When the wood is thoroughly kindled, add more coal. A blue flame*

---

16    The Original Fanny Farmer 1896 Cookbook The Boston Cooking School A facsimile of the first edition published in 1896 Ottenheimer Publishers, Inc. Commemorative edition published in 1996. P 19-21. This Company closed in 2002.

*will soon appear, the gas CO* (carbon monoxide) *in the coal burning to carbon dioxide. When the blue flame changes to a white flame, then the oven damper should be closed. In a few moments, the front damper may be nearly closed, leaving space to admit sufficient oxygen to feed the fire. It is sometimes forgotten that oxygen is necessary to keep the fire burning soon as the coal is well ignited, half close the chimney damper unless the draft be very poor.* [Carbon monoxide was a silent killer.]

*Allow the firebox to be no more than three fourths filled. When full, the draft is checked, a larger amount of fuel is consumed. This is a point that should be impressed on the mind of the cook.* [You don't want to waste your resources. You want to heat the food and the room not the chimney.]

*Ashes must be removed and sifted daily; pick over and save good coals, –which are known as cinders– throwing out useless pieces, known as clinkers.* [Remember the nursery rhyme 'Little Tommy Tinker.']

*If the fire is used constantly during the day, replenish coal frequently, but in small quantities. If for any length of time, the fire is not needed, open check, the dampers being closed; when again wanted for use, close check, open the front damper and with the poker rake out ashes from under fire and wait for fire to burn brightly before adding new coal.*

*Coal, when red hot has parted with most of its heat. Some refuse to believe this and insist upon keeping dampers open until most of the heat has escaped into the chimney* [wasting fuel and this could possibly lead to the danger of a chimney fire.]

*To keep a fire overnight, remove the ashes from under the fire, put on enough coal to fill the box, Close the dampers and lift the back covers enough to admit air. This is better than lifting the covers over the firebox and prevents poisonous gases entering the room.*

Portable kitchen cabinets that were common in pioneer homes were not much wider than four feet with an upper and lower cabinet, a metal lined bin for storing flour and space for cooking supplies and several drawers for small tools. Limited counter space. Whatever they required, they learned to 'make do' with what they had.

*Remains of an old fashioned portable kitchen cabinet at the abandoned Gronvold homestead, from a picture I took on my 2014 Family History Trip.*

*John's barn. 1982, towards the end of the barn's life, when the Gronvold cousins from Seattle were visiting. Clarence is in the centre in dark clothing. Alvin Gronvold is the tallest man on left and Walter is the man with the beard over to the right*

\*\*\*

JB said *"I built a fairly big barn west of the house near the well."* That was around 1915. Lean-to additions were later built on both the north and south sides.

*One year in the threshing, Dina drove a load with two horses, and a little boy, Tommy Dunn, drove a load behind, and they made a trip a day, with a little over 100 bushels on the two loads. They hauled in 1000 bushels before I got to town.*

(I wished J.B. had put a date on this event but I believe this was in the very early years. Tommy Dunn was the son of James Dunn, who lived about three miles west of the Dahl Melby place near Beaubier, but I understood this family may have lived west of the Hansen place in the earliest years.)

The grain hauling process was explained by my cousin Larry Hansen, in an August 2017 email:

*"They would shovel a load of grain from the wide door granaries at JB's farm – all by hand, there weren't grain augers at that time.*

*At first light, they would be harnessed up and leave for the elevator at Tribune* [or Colgate if it was before 1913].

- *Once in the elevator they would unhitch the team from the wagon.*
- *Then, a lift, powered by the old Fairbanks-Morse elevator engine,* [with its distinctive 'putt-putt, putt-putt' engine], *raised the front of the wagon.*
- *A slide door, at the back of the wagon, was lifted to allow the grain to be emptied*
- *Shovel out the corners,*
- *Lower the front of the wagon back down,*
- *Hitch the team up,*
- *Go to Lee Hing's Chinese Cafe and have a complete dinner for 10 cents,*
- *Then head for home,*
- *Back up to the granary,*

– *Shovel on another 50 bushels and by that time, it would be getting dark.*

*The next morning, they would hitch up and be gone again. If they hauled 1000 bushels to Tribune, it would have taken 10 days with 2 wagons. There weren't any established roads at that time. Bernhard told me the trail went north from the Dravland School, crossed Long Creek by the Anton Kuntz place, and then continued northeast to Tribune."*

***

My Grandmother on horseback, or helping out in threshing, was not something I expected. She would have been busy enough with caring for the family, but I'm not surprised. There was likely one of the Johnson family, as a hired girl, to look after the little ones during those busy times. And I am sure, the ease Dina had with horses, encouraged my mother's love of horses though we never did have horses on the Melby farm.

***

There were 'ne'er do well' types, who made things tougher for others. JB described this situation, "*I think it was in 1914, a man came to buy seed grain from me. When he pulled out big rolls of money, I thought he was a great guy, but the opposite proved true.*

*Another day, when I was plowing with 6 horses, he asked me to go to town with him in his fine big car. He went to the bank and asked to borrow money. Lucky for me, the banker refused since he was on the blacklist---or I would have signed for him. I [had] hauled out lumber for him for a big house, gave him various furnishings and threshed his crop, so all told, I had $500 coming from him, and got nothing."*

*John on the disc with six horses. When the pictures match the commentary, it makes the story come alive.*

\*\*\*

Now the two daughters in the family are joined by two sons. This next part of the story was my most interesting discovery.

*When our family increased with Bernhard, born Oct 30, 1913, and Clarence, May 8, 1915, we had to build a 12x12 sleeping room onto our shack.*

At first, I thought this picture, from my grandmother's postcard collection, was an anonymous postcard of a threshing crew in operation. I originally intended to use it as an illustration back in Chapter Two, for the time when John and his brother, Adolph were in Portal, ND for jobs as harvesting crew. But when I shared it with my cousin, Larry Hansen, he recognized it as an early picture of the Hansen homestead house in the background. My uncle Bernhard had shared with Larry that the high side of the house was facing south away from the northwest wind. The barn would have been built to the right of where the hayrack is in the picture.

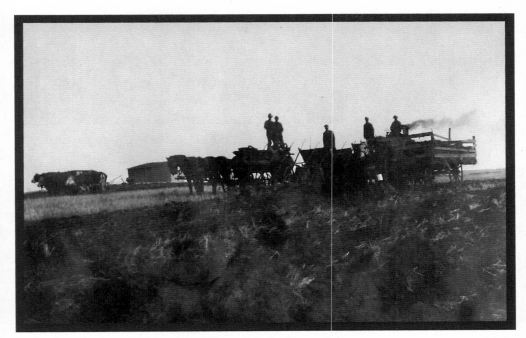

*Silhouette of a breaking crew.*

Using a magnifying glass to study the image, it clearly showed the typical tar paper shack- where black tar paper is placed over the wooden walls and is fastened down with perpendicular lathes. The new "12x12 sleeping room" addition, John described above, is the section on the front, gleaming brighter in the sunlight for it is built with lighter horizontal boards. There's a window in the middle of that wall, difficult to make out, as it is seen through the horse's harness.

Larry[17] recalled from grandpa's story: "Olaf had bought a breaking machine that soon turned over the rest of his homestead. It is pulled by a steam tractor and is an 8 or 9 bottom breaking plow. Our Grandpa had one like that. Those are the lifting handles that you thought were pitchforks." You can see the stubble has been freshly turned over in the foreground.

---

17     Email from Larry Hansen Jan 4, 2018, and corrections March 13, 2021

Two men are posing on the water tank, two on the hayrack and one on the steam engine or tractor.

There is a picture of a breaking plow in Chapter Four, likely the same machine shown in this picture. "It has been used to plow a fairly wide band around the dry stubble. The black and white team on the right side is with the straw rack, where the sheaves of grain are thrown into the threshing machine or more likely here, straw in this case, is thrown in to provide fuel for the steam engine—and on the left side is a team on the water tank, to fill the steamer with water. Both straw and water were filled several times a day. The steamer could be fired with straw if coal wasn't available."

\*\*\*

In the background, a team of three or four oxen are in the process of ploughing a narrower furrow around the house to act as a fireguard. I thought surely one of these must be his faithful ox, Buff, that he mentioned earlier. The photographer is on the northwest side of the buildings facing towards the southeast, making the picture backlit. My best guess as to location would be north of where the barn was eventually built approximately where the original prairie lane came up into the yard. And who of the five men was my grandfather, I guessed it might be the man on the far right or the second from the left based on the type of hat he'd been seen wearing in other pictures of that time. The other men might have been John's brothers-in-law, Olaf and Adolph and father-in-law, Johan Gronvold, and possibly Ray or Jesse Erickson who'd been mentioned as assisting John in operating the steam engine, for Ray had been trained as a stationary engineer.

\*\*\*

*Hansen family with four children, Edith, Myrtle, Bernhard and Clarence. Fall, 1916. This was the last formal family photograph until the mid-thirties.*

\*\*\*

The 1916 census offers this consideration of the family,[18] John is 38, Dina- 28, Edith- 6, Myrtle- 4 Bernhard- 2 and Clarence-1. In transcribing this document, the youngest child's name is given as Clarence Aldio, when his name is, Clarence Alvin. In the original census transcript, I saw that name had been written twice and scratched out. I added this footnote[19] to the item

---

18    Year: *1916*; Census Place: *Saskatchewan, Weyburn, 02*; Roll: *T-21946*; Page: *18*; Family No: *216*

19    http://.ancestry.ca/cgibin/sse.dll?gss=angsg&new=1&rank=1&msT=1&gsfn=Cl arence+Alvin&gsfn_x=0&gsln=Hansen&gsln_

"I believe the census taker, Reg Lauder (who was the grandfather of my school friend, Donna Rososki) likely had trouble understanding their heavy Norwegian accent."

***

*I started to raise pigs and had quite a few. When feed got scarce and pork prices low, I got Leonard Johnson in to butcher them when quite small and froze them* [in Mother Nature's deep freeze] *to eat that winter.*

*We didn't do too bad raising cattle. We hauled quite a few to Ole Stovern, the butcher in Tribune, and prices were fairly good.*

*The Hansen family (about 1920) visiting at the Axel Erixon place in the Summit district, south of where the Hansens lived. John had the adjoining quarter of land next to them that he used for pasture for his cattle. Children, l-r Myrtle, Bernhard, Clara, Anna, Clarence, Edith. It's interesting to note that all of the children's clothing appear to be made from the same fabric.*

## Using all but the squeal

Side pork was commonly put in a brine to make salt pork. Curing it this way was a simple system to keep meat from spoiling. Fried salt pork, accompanied by a cream gravy made from the drippings along with boiled potatoes and canned vegetables was a favorite easy meal, a common supper meal I remember at our house in the years before rural electrification.

Another common way to preserve meat was by canning it.

The fat from the pork was an important resource, used in several ways. First of all, the excess fat, trimmed off the carcass, was rendered to produce lard for baking or frying needs. Once all the fat was melted out, cracklings were left – a crunchy snack item enjoyed by most.

And even after the fat had been used too many times for deep frying or the cook had saved the excess pan drippings after cooking pieces of side pork, it still did not get discarded. That grease was saved to be used to make soap.

Lye, needed to make soap, could be made by leaching out the lye in ashes left from burning wood and coal, but I remembered boxes of Gillette's lye on the back shelf from when I was very young. I was aware of the fact; homemade soap was still being made and used.

A non-corrosive pot filled with the fat, lye and water was placed over low heat at the back of the stove, and several hours later, when the chemicals had interacted, the liquid soap was poured into a mold of some kind. Then it was left to set and get dry. Soap needed for laundry would be grated with a medium fine grater, the same as you would grate a hard cheese, like parmesan. And those bars of soap were used for all household and personal cleaning purposes. At some point, there may have been products purchased from a J.R. Watkins or the Fuller Brush Travelling Salesman.

Can't forget one more item. Some, like my mother, considered pickled pig's feet, a delicacy, but not me!

# Now we made soap, let's do the laundry!

Doing laundry was a big production. It was hard physical work compared to what's involved today. Monday was the traditional wash day. In winter, large copper boilers and other tubs, filled with snow the night before, were set on a heat register or on the stove to melt and heat up. I experienced doing that when I was growing up. But because the well on the Hansen farm had a windmill to pump the soft water, they were some of the fortunate farm families who didn't have to conserve every bit of water that they could scrounge whether from a cistern, rain barrel, snowbank or conventional well. They likely were able to skip that step.

Before washing machines with mechanical agitators to swish the clothes in the soapy water became common, a washboard was used to scrub dirty clothes. Without a doubt, an item in the homestead shack, was a washboard –with its knobby ridged glass- set in a wooden frame. Scrubbing the wet garment, up and down over the ridges helped to loosen up the dirt and stains with mechanical action, but it was a little tough on your knuckles and harsh on the skin. Then, unless you had a mangle to squeeze the excess soapy water out of the clothes, you still had to twist and wring them out by hand before rinsing them in clear water. Then, the wet laundry would be hung onto several clotheslines out in the backyard, using clothes pins.

Tuesday was considered the day to do ironing, and since 'wash and wear' is a more recent innovation, then most washed items were also ironed. But first the dry clothes would be sprinkled slightly to help remove the wrinkles once the heat of the iron was applied, as a steam iron does today. Before we had electricity, we had several flat irons (or also called sad irons)-one to use and one to reheat, which you did by placing it on a hot stove top to heat up. You checked the heat by licking your finger and

> tapping the bottom of the iron. If it sizzled, the iron was hot enough to do the job.
>
> Considering all the physical work involved in laundry, family members had to be careful to keep their clothes clean, so they could last for the week. Each individual family member did not have many changes of clothes, like most of us do today.
>
> I remember coming home from school on a winter wash day about 1955, when my job was to bring in the long johns, sheets and towels from the clothesline in the side yard. Of course, they were frozen stiff and had to be draped over fold-out wooden racks to finish drying. Jostling an adult sized pair of men's long underwear along with five graduated sizes of boys long johns create a funny image. (Remember, I was the oldest of the family and I had five younger brothers.)
>
> But there was nothing like the scent of all outdoors on those clothes. Downy or Fleecy doesn't begin to compare.

Long before we had television, my brothers and I spent that time, between returning from school until supper, listening to a radio play. **The Lone Ranger** was one of our favorites.

Or I'd take down my mother's photo albums and study them. I remember feeling impressed and often a little jealous for studying my mother, Edith's photographs showed a variety of groups who always seemed to be having so much fun together. Later, I'll reveal some of Edith and her siblings' experiences in their growing up and young adult years. They are not part of John's memoir, but they do help to fill out the details of the family story.

*** 

Since Lutheranism was the national religion in Norway most Norwegian immigrants were Lutheran. The spiritual aspects of their lives were important. Now this next section showed how they put it into practice. I know it was their faith that saw them through the many trials and struggles they faced from the early to later years. I prize the rich Christian heritage their lives demonstrated. Though Christianity is being challenged in many arenas of life today, we can learn and be heartened by their example for, as believers, we are part of that 'communion of saints'! The experiences of the saints before us can be an encouragement to us when we face similar challenges.

***

CHAPTER 6

# THE COMMUNION OF SAINTS

*Church history in the homesteading community*
*1909-1960*

I was fascinated to imagine the group of people gathering for a December 27, 1909, meeting to organize the first Lutheran congregation in this area when I learned about that historical fact decades later. Back then, it was the Chris Andersen home; four and five decades later, the house where I grew up.

A timeworn scribbler stashed in my mother's file cabinet when she lived in Trinity Tower in Estevan was filled with Norwegian notes in JB Hansen's distinctive handwriting. They turned out to be the minutes of the Hauge Lutheran Congregation. The charter members listed included both sides of my family, my maternal grandparents, Dina and John Hansen, as well as my great grandparents (Dina's parents), Johan and Maria Gronvold and my paternal grandparents, Dahl and Elise Melby, Elise's father and my great grandfather, Jens Hansen and my other set of great grandparents, Dahl's parents, Lauritz and Karen Melby. All of the Melby group lived about 10 miles west. But the rest of these folks were from the local area- Chris Anderson, J. Domstad, O. Saxhaug, Andrew Julstad, G. Rasmussen, O. Lokken, H.O.Fossum, E. Kelstrum, Leonard Johnson and Bernhard Johnson.

*This gathering could have been a church service or a Ladies Aid (the name given to the church ladies group) meeting at the end of the 20s or early 30s, as some of these folks moved up to northern Saskatchewan in the 30s. You'll encounter these names later when we come across Edith's trip diary of 1933 talking about the Luther League Convention at Birch Hills.*

Edith listed these names on the back of this picture: Saxhaug, Lokken, Ole Torkelson, Parnas, Fossum, Mrs. Domstad, Alice [Lokken] and [her new husband] Obert Johnson at the J.B. Hansen home.

Unfortunately, that humble notebook went missing, so likely the only accounts of Hauge Lutheran Church are what I record here, though my mother did use this information almost fifty years ago, when she wrote an account about the local Lutheran congregation[20] on page 119 of the original community history book: **The Saga of Souris Valley R.M. No. 7.**

***

---

20      **The Saga of Souris Valley No. 7** Friesen Printers (a history of residents from 1906-1976) p. 119

I had wondered but I found that knowing about why they picked the name, Hauge, helped me to better understand my grandfather's unwavering beliefs and practices.

Once I learned to read, I scanned absolutely every newspaper, magazine or letter that came in the twice weekly mail. We used to get the **Hauge Innermission** paper, as did my grandpa J.B. I knew he was active in events they held. I realized the articles matched up with what my grandfather practiced and spoke about.

Grandfather's pietism and legalism affected me. I was at the wedding of a school friend, in Estevan. The bride and six other local girls, including myself, had been part of a small choir, led by the United Church Student Minister in Oungre, over several spring and summer seasons. We'd go along with him on a few occasions to the various points in his charge- Round Grove, Oungre and Bromhead and sing several special songs for the Sunday worship service. With at least four of us crowded together in the back seat with our gathered border print skirts and the crinolines of that time, this was one part of the fun of this opportunity for voice training.

But this time, because of our special association we were asked to assist with serving at one of the girl's wedding receptions at Oungre. My friends usually wore lipstick, and this time I did, too, for I had found a lipstick sample among some old things of my mothers.

Grandpa happened to be there at that wedding ceremony in Estevan because the groom was the son of another early homesteader. And besides, he didn't need an invitation to attend a church wedding service and it was just down the street from his house. I had spoken to him only briefly as we were leaving from the wedding.

It took me some time to come to terms with this letter I received from him several weeks after I was settling in at boarding school for my Grade Twelve year at LCBI, Outlook. He had written, *Dear Granddaughter, Pardon me, but I figure it was my duty, to let you know I was more than a little disappointed....to see you with such red lips – Yes, Elaine, you are hereby telling God that he did not make you in right colour so I honestly think that you see fit to quit with that sinful habit, especially now that you are to go to Outlook, that you should not go there to continue ruin yourselves and also other students with your bad example. So, Elaine, I will say, what I said to Clarence when he had got the habit to smoke in the army: Quit it and you are the one that gains. And he did quit, which was a good thing, being as now, he has three boys. I also appreciated it very much that your father did quit smoking. Hope you will be able to quit this sinful habit and throw the lip paint so far, you don't find it.*

*May God help you to overcome and be an example also for your younger girl cousins, as they will probably look to you to see* [what] *the oldest is doing.*

*From your grandpa*

That was probably the first time I had ever used lipstick. As I was growing up, Mum's 'beauty routine' was never anything more than Jergens lotion and a quick fluff of face powder. But, as busy as her life was with six children, she likely didn't have time for any more than that. And no (or minimal) makeup does make life simpler. Once, when I was reading a book about skin care and makeup, my father's comment brushed it off as, 'just foolishness.' But occasionally, I try and improve upon nature. After all, I could recognize even in black and white photos that Billy Graham's wife was wearing lipstick!

\*\*\*

We were not allowed to go to movies, although we enjoyed movie nights at the community hall sponsored by some agricultural companies (some of them were National Film Board productions) or movies re-enacting the life of Christ at the church. The first and only time I was in a movie theatre when I was growing up, was to see the movie based on the story of Martin Luther, during the time that I was studying for confirmation.

Once, after spending a Saturday playing with one of my Jewish friends, her family invited me to go along with them to a movie being shown at Torquay. Permission was requested, but my father would not let me go for we had to get ready for Sunday.

Or when the school wanted us to take part in a square-dancing activity when I was in Grade Three or Four, we were not allowed to take part. One day, when I was in Grade Six, the furnace in the corner was not working and the classroom was cold, so, a record was put on the record player and we all joined in a conga line, doing the Bunny Hop around the room. It was fun, great exercise, we got warmed up, but somehow, I still felt guilty!

And when my uncles first brought their prospective mates to meet and to visit at my grandparents, they were warned to make sure they had "sensible shoes" along and to change out of their high heeled pumps and/or remove earrings if they wore them.

But it's not about what we do- that's the Law. It's about what Christ did for us, the Good News of the Gospel as summarized by John 3:16. "For God so loved the world that he gave his only begotten Son that whosoever believes in Him shall not perish but have eternal life."

***

In the late fifties, my father got a book, from my aunt Myrtle, in the family Christmas gift draw, which told the story of Hans Nielsen Hauge. That started my research on this charismatic lay preacher in Norway.

Hauge experienced what has been commonly described as his 'spiritual breakthrough' on the 5th of April 1796 while he was out ploughing a field on his father's farm in the Tune, Ostfold area of southeastern Norway, (where the Melby family were from.) The 25-year-old considered himself to be a Christian, he had been baptized and confirmed, and he read the Bible and devotional books daily. So, as opposed to this experience being an act of repentance for an immoral life, it was more a deeper commitment to and an experience of the divine that introduced a radical new perspective into his heart and mind.

Compelled to share the salvation story, Hauge travelled by foot over the length of Norway preaching about "the living faith" at countless revival meetings which got him in trouble with the authorities because he was not an ordained clergyman. "At that time Norwegians did not have the right of religious assembly without a Church of Norway minister present. The time he spent in prison broke his health leading to his premature death."²¹

Nonetheless, Hauge, is revered as a leader in Norway, credited with starting farmer's movements, folk schools, running successful businesses as well as writing and publishing many books.

## Hauge

Since I recognized such a strong Haugean influence on both sides of my family, I discussed matters relating to Hauge at length with Tim Peterson, who is married to my Melby cousin, Donadee. Tim is a Lutheran Pastor, and has been in Norway, so I valued these insights he provided for me in a February 2017 email:

*The Haugean people built the Bedehus (translated prayer house) in many areas to supplement their church involvement. They still attended worship at church but had extra meetings* [like prayer meetings and Bible study] *at the Bedehus.*

*I am fascinated with Hans Nielsen Hauge and the Haugean movement. I read a biography of Hauge when I was quite young. I suppose my parents encouraged me to do that.* Harold Melby (Tim's father-in-law and my uncle) *had a final major paper he wrote about Hauge, before he graduated from seminary.* (They thought no copies had survived, until an inquiry to the Lutheran Seminary in Saskatoon did turn up Harold's paper.)²²

*For me, personally, Hauge is a hero of the faith - Peter, Paul, Stephen, St. Augustine, Martin Luther, and Hans Nielsen Hauge - God used them all to bring the message of Jesus to His People.*

---

21    https://en.m.wikipedia.org/wiki/Hans_Nielsen_Hauge
22    email Feb 15, 2017, and Feb 28, 2021, from Tim and Donadee Peterson

*Many of the Norwegian immigrants (including my family, as well as your family) were very much influenced by Hauge. The focus of his preaching was to have a living faith in Jesus and to have one's life changed by that faith. He believed that our lives should be changed as we look to Jesus and follow him, but he was not legalistic in the way his followers became. That's obvious regarding alcohol. He spoke out strongly against drunkenness but was not opposed to all use of alcohol. That came later.*

*In my opinion, the legalistic emphases among the Norwegian Pietists had good reasons initially, but they were sometimes disconnected from their original purpose. Alcoholism is an obvious problem and one solution to that problem is to ban alcohol entirely. Card playing was originally much more connected to gambling. Dancing was seen to be connected to improper contact between men and women. The concern about jewelry and clothes is connected to certain Bible passages that talk about the proper adorning of a woman. When the Pietists wanted to live their lives completely for God, they wanted to totally reject these things.*

*The Pietists were very convinced that certain things were wrong. It can be hard to understand some of their views, but I think it can be helpful to remember that their main concern was always to live a life that was faithful to Jesus.*

"Carrying on the Hauge tradition" in the present day, Tim took on the challenge of fund-raising efforts in the USA to preserve the museum at Hauge's childhood home in Norway. He also helped establish an annual Hauge lecture at the Norwegian Lutheran Memorial Church in Minneapolis.[23]

\*\*\*

Dahl Melby, my paternal grandfather had attended several years of seminary at Red Wing, Minnesota. I learned that school followed the Haugean and Pietistic practices of experienced conversion, and simplicity in worship, when I researched the story

---

23   From a Feb 28, 2021, email

of my maternal great grandmother Maria's sister, Bertha Espe. There were some connected to her husband's family who were instructors at this school. Dahl Melby somehow wasn't able to complete his course, so he wasn't able to be ordained as a pastor, though he often used his training by serving as a lay preacher in the various districts he lived.

***

In the early years, the church records showed concern with developing a cemetery[24], as well as paying the minister. This congregation was going to pay $75 a year for the pastor's services. A motion from Jan 2, 1913 *meeting indicated they planned to hold 13 services a year.* [That would be about one a month.]

Later in 1913, when the group further west (including the Dahl Melby's) were forming a new congregation closer to their home, the Melbys were released from their responsibilities here. The notation "*Dod*" (Norwegian for died) was next to the names of Karen and Laurits Melby, on the member list. They died within months of each other in 1912.

The Annual Meeting minutes of January 1914 recorded *that the secretary* (that was JB Hansen) *was to write to D.L. Melby to ask him to do the job of going and collecting for the pastor's salary, now listed as $125.* It appeared that rather than passing a collection plate, a layman from the church took the responsibility of going around to collect dues, also as happened in another area church, according to the history of the Nordalen Congregation, northeast of Lake Alma[25].

---

24    A note in the Cemeteries section of the 2005 update of the community history, **The Continuing Saga of Souris Valley** specifies the cemetery is located 5 miles west of Oungre and two miles south of Highway 18, on land that was donated by Bernhard and Bina Johnson, whose homestead was nearby.) This cemetery is where all of the Hansen family, as well as Johan and Maria Gronvold and Adolph Gronvold are buried.

25    Lake Alma Over 50 Club **Settlers of the Hills** D W Friesen 1975 page 123

## Ladies Aid

And we must not forget the important role "Ladies Aid" played in ensuring that the minister got paid. It was also an important social event. On the occasion when the Hansen's neighbor, Mrs. Leonard Johnson "had hosting her first Ladies Aid, she had a full house. Some ladies walked, others came by horse and buggy."[26] Often the men came along for the food.

The organization in my day was called Lutheran Church Women- it was a group of church women that met for Bible study, fundraising, food and doing things that needed to be done in the church.

When I was growing up, a box in our basement held a set of four dozen blue edged white enamel metal plates. They came out on the rare occasion our family had a picnic, like at the special Sunday service at Midale Lutheran Bible Camp or a family picnic at Woodlawn Park in Estevan with our grandparents and some of our aunts and uncles. They were the 'Travelling Ladies Aid Plates.' No one would have had enough dishes to serve a crowd and I'm sure there weren't paper plates or enough spare cash to buy them if there were. At what point they were obtained, I don't know, but I managed to keep one for that history's sake.

---

26   **The Saga of Souris Valley** p115 from the Leonard Johnson story written by Victoria and Mable

*Community gathering at Melby's.*

A community gathering in the spring or summer of 1911 at the Dahl Melby place.

The Melby house had been a store building moved from Ambrose, and while not large, was considerably bigger than the typical one-roomed shacks many early settlers had.

The people identified, starting from the middle row of adults, from left to right: Bill Keller (with white tie, Melby's neighbor to the south), ??, Henry Hansen (with arms crossed was my grandmother, Elise Melby's brother), woman behind ??, Ole Torkelson and his new bride Selma; the couple in dark clothes are Mr. &Mrs. Henry Fossum, between them, partly hidden is Harold Torkelson, and woman to the right of Fossums is Rena, Harold's wife, holding Agnes, and just to the front of her, with the white V neck, is Mrs. Bernhard Johnson. Peeking around to the right of her head is Dina and John B. Hansen, in a sailor hat, holding toddler, Edith, then two unidentified women holding toddlers, Elise Melby, holding Victor in white shirt and dark overalls, a woman ?? in dark dress with a child; above her, the tallest man is Emil Torkelson, holding his young daughter, Lillian, and Hulda, his wife, partly hidden beside him, Dahl Melby with the dark tie, Ole Torkelson Senior, Bernhard Johnson, Jens Hansen and Swan Petersen. (*Jens Hansen wass Elise's father, Jens's sister, Maren, had been married to Ole Torkelson, both their wives had died back in central Minnesota. So Ole, Emil and Harald Torkelson were Elise's cousins*) The others at the back are not identified except for John Claybo, the man in a hat in front of the right hand window. Children- l-r Alvin Torkelson, Orville Fossum, and the little girls were said to be from the Harold Torkelson family, and that could only be Hazel and Bessie (but there are four smaller girls, all wearing similar dresses, the tallest girl is Anna Melby and her brothers, Elvin and Daniel are in the white shirts with Morris Torkelson between them.

Identification was made by Edith Hansen Melby, Ken Torkelson and me, based on family resemblances and the process of elimination. For example, on studying the photo with a magnifying glass, four girls are wearing light print dresses, all made from the same material. This made me think, they might be from the Eric Elving Family, who did have five daughters and lived close by.

Most people in the picture are young though I thought there were some elders who should also be on the picture, like Laurits and Karen Melby, my paternal great grandparents, but perhaps they are some of the indistinct people in the back. My guess, Karen is barely visible, standing in the doorway of the house and Lauritz is the second man over to the left of Emil Torkelson with his hat, partly covering his face, directly behind Elise Melby. Records showed Dahl Melby had a brother, Mike who homesteaded in the early years but returned to the States within a short time. And then, when you see the names of the people involved with the Northern Light Society events, described below, these are likely some of the other unidentified people.

***

This picture was *assumed to be a gathering for a church service, Dahl probably conducted the service. I can remember Dahl speaking in church as a layman,* Ken Torkelson told me.[27]

But it could also have been a meeting of the **Northern Light Society,** a Scandinavian Young People's Society[28] organized in those early years. At least two of these meetings recorded took place at close by neighbours of the Hansen's. The meetings were most interesting. They even held debates.

*"Feb 18, 1912, at D.L. Melby. A great crowd present. Ladies Aid served the lunch. Our society to be named "Northern Light" and our newspaper to be named "Forward." Our contribution and also the Laws, Articles and Dues to be decided later. Carried. Mrs. Ole Torkelson elected Treasurer.* I wondered if that could have been when the picture was taken but the only child with the Hansen's is my mother and by this date Myrtle, the second-born would have been almost four months old.

---

27    In a letter from Ken Torkelson sent to me, dated Feb 5, 1998, when I was writing the Melby family history for the **Settlers of the Hills** community history of the Rural Municipality of Lake Alma.

28 -    This material documenting the Northern Light Society was found at the Lutheran Parsonage in Lake Alma when my sisters- in- law, Priscilla and Sheila Melby were researching for the "Melby Family History" in the late 1980s.

*[Date not distinguishable]-* At John Skjonsby. The meeting was called to order by D.L. Melby. The program was good especially the debate, "that money was better than a good name." Frank Mathson was there with the goods.

Next meeting at F. Mathson. We all should bring our lunch and our collection should go to songbooks.

*Aug 4, 1912-* Meeting was held at home of Mr. and Mrs. B.J. Johnson [neighbours west of Hansens] and was called to order by the president. A song by the audience was first on the program, followed by Scripture reading by D.L. Melby.

The following officers were elected for the next six months: President- D.L. Melby; Vice president- Fred Dravland; Secretary-treasurer- F. Mathson;

Program Committee- R. Sovig, Lillie Peterson, Serine Hanson, Fred Dravland and John Dravland.

The following program was rendered:

Reading: Fred Dravland

Song: Choir

Vocal Solo: Berger Claybo

Debate: "That it is better to live in the woods than the prairies."

Affirmative: B.J. Johnson, F. Dravland and D.L. Melby

Negative: John Skjonsby, J. Claybo and Theo. Sovig

The judge's decision was 2 to 3 in favor of the Negative.

Reading: Ben King (not present)

Song: Choir

Editor: John Claybo

Song: D.L. Melby

Critic: F. Mathson

*Nineteen new members were added to the society. Lunch was served after which a few games were played by some of the young people. The program committee made out the program for the next meeting.*

The next part of the records included meetings at several places, with fewer details about the program:

*At Emil Torkelson with a "debate concerning the Negroes, Indians and the White man.*

*B.J. Johnson- Bad weather that day not very many present.*

*Fred Dravland- a big crowd was present......Ballgame, running races of the long and short kind.... all enjoyed themselves immensely.*

*Swan Peterson- Called to order by Vice President Ed Jacobsen....The lunch were dandy, and all had a good old time. June 16, 1912*

*R.J. Sovig- the biggest crowd was there that so far had come to our Young People's Society. The day was very nice, so the meeting was held outside.... Program good, especially Mr. Barnett's speech* (he was the teacher) *...and also Mr. King had a good speech. Picture was taken.*

My hunch was that the organization of the St John Congregation further west may have taken the place of the Northern Lights Society. I knew that in 1914 the Dahl Melby family moved to Hillsboro, in southeastern North Dakota for a season or a year or so. Since Dahl was attributed as the leader, that also may have had something to do with it.

*\*\*\**

Now we switch gears away from the 'more active than I'd ever imagined' early church, social and community life and discover the JB Hansen's homestead shack is bursting at the seams.

*\*\*\**

CHAPTER 7
# FAMILY EXPANSION

*Of houses and horses*
*1919–1930*

John's memoir now details the move into a new house, and we learn more about horses and farming: *We had three more children, Anna born July 24, 1917* [named for her aunt] *and Clara[29] Lillian, June 18, 1919 and Evelyn Violet born May 21, 1921.*

*Now it was necessary to build a bigger house, as we also needed a place to board teachers* [as well as a hired girl or hired man, was a note my mother, Edith, added]. *The new house was finished in 1919 and cost only $3500. We did much of the work ourselves and we were able to buy lumber for a set price, quite cheap. I took a loan on the land for $3000. I thought it wouldn't take too long to pay for it, but when crops were poor, I didn't get it paid for until 1943, but we had a good house all those years, for which, we thank God.*

\*\*\*

Once the house was completed, the community collected money for a gift: a set of Mission style side chairs that were presented to the family at the housewarming party my grandmother had told me about. My mother

---

29   Clara's namesake was Dina's 'special cousin' and friend, Clara Rude, who was the same age as Dina and died shortly after the Gronvolds moved to Ambrose

reupholstered the rocking chair with a piece of her own crewel embroidery work. After my second son, Nathaniel, was born in 1982, she gave it to me.

The side chair must have been discarded. In the family group picture[30] from Maria and Johan's Golden Wedding anniversary celebration held in 1925, Maria and Johan, Dina's parents are seated in those chairs. And the chairs show up in other photos taken on the south side of the house over the years, including when I sat in it at the cabin my grandparents had at Manitou Lake.

There was an old-fashioned pump organ in that living room, until the early fifties. When Clarence was married and took over the home place, Vera's piano replaced it. That organ came to the Melby home and was where I first learned basic keyboard skills. That experience came in handy, for there were times, I was the only one available to play for the hymns at church- on an old pump organ. (We did get a piano later.)

***

What happened to the old homestead shack after the new house was built?

I asked my uncle Johnny, who said, via a January 2017 email, from his daughter, Rhonda Trueman, "The old house was used as a granary, after they moved into the big house. They had a feed grinder in the smaller part of the building where they ground the grain, usually barley or oats. The chop [*ground grain used as a supplement for the cattle and horses*] was stored in the other part."

Though I remembered that little building on the south side of the barn; I had never realized this had been their homestead. I asked if anyone knew if any pictures existed of that old homestead shack. And my cousin Larry said he had a photograph of part of that building taken before it was demolished. Some weathered wood saved from the demolition was used for the red frame of that picture. He also said that Bernhard had shared with him that the original homestead location was at the southeast corner of their new house. That picture aside, I was destined to find the picture I wanted to round out this story, as you should have already discovered.

---

30    The Princess Doll's Scrapbook Page 46

Before the days of running water and flush toilets, there was another functional building, tucked up close against the caragana hedge, northwest of the house. It was painted a creamy-yellow with dark green trim to match the house. I always thought the toilet (aka outhouse) at the Hansen Place was grand, at least compared to the small weathered gray structure built from recycled wood at our farm. Its only delights were the shafts of sunlight coming through multiple nail holes, playing on the dust motes in the air. That observation could prove a fascinating study, while trying to while away some time, to get out of, or at least delay doing the dishes.

This one at the Hansen homestead had three holes, with one custom sized in height and perimeter for a younger child's bottom. A wooden box or shelf in the corner left of the door, held a stack of old newspapers and catalogues; and in some seasons, like at Christmas or fruit season, coloured tissue paper wrappers that had been carefully flattened out and saved for another purpose. You might wonder what they did in the cold of winter. That was when a honey bucket, as they were euphemistically referred to, came into use in one of the closets in one of the four upstairs bedrooms. They could teach us a thing or two about Reduce, Reuse and Recycle. But I still like my roll of Charmin.

*We had started with pigs again and had about 300 which sure helped pay expenses.*

But based on his earliest experiences, John continued- *I did better raising horses, since I got $400-$500 per team for full-grown horses,* except in this circumstance, where he said, *I bought a team on an auction sale for $425 and gave a note till fall. A letter from the Bank of Montreal said it must be paid in full. I couldn't pay more than $100 and sent it, shortly after the man went broke, so I lost the $100, so it got to be an expensive team.*

*We had good luck raising horses when I needed more for the larger machinery. We used up to 12 horses, six on each outfit,* John said in another place.

*Yes, the working days were long,* John admits, as he explained his typical workday chores. *As a rule, I was up first, got the fire started in the stove, and then went to the barn to feed the horses and curry and brush them before breakfast. When we'd eaten, two of us had to get hitched up, so we could get out to work at 7 o'clock. At 12, we unhitched, ate dinner and then went out to feed the horses oats, and as a rule, we had to oil up and see that all the machinery was in good order. Then, we hitched up again at two and worked till seven. When we put the horses in, we unharnessed them and gave them hay. We ate supper and then went out again to give the horse oats and curry and brush them again. We filled the manger with hay for the night and bedded them down with straw.* John took good care of his horses.

### Edith's horse story:[31]

*I loved horses and spent much of my time on horseback— not just for fun either— for it soon became my job to gather in the horses and cattle who were allowed to roam the prairie and find feed, to see that they were OK, and maybe chase some of them home. Our pony was a sorrel named Nelly, the same age I was, [who] gave many rides to novice and expert, treating them accordingly.*

When I first researched my original 'doll story', I discovered Nelly had been an older twin cousin of my grandmothers, who died as a young woman after the Gronvolds moved from Fertile to Ambrose, ND. I'd guess, my grandmother, Dina, named this pony.

*Nelly was the pet of the family but if a new beginner got on her back, she would seldom run. When she died at an old age, Dad had her skin made into a robe much needed on long winter drives. But the tanners got the hides mixed up and we received a bearskin robe. When the*

---

31   "The Family Tree Scrapbook" A personal family history from my mother, Edith Hansen Melby, filled out as a gift for my oldest son, William Ayre.

*robe was used, the horses detected the bear smell, and they almost had a runaway, so exit bear skin robe!*

*Edith on Nelly The hedge that appears in the background was likely the start of the grove of trees that surrounded the farm on three sides.)*

\*\*\*

*Dad had a good voice and after weekends, when his work horses were let out on the Hudson Bay Land,[32] north of us, he would call each horse by name and home they'd come to get their oats and to get harnessed for work. Father raised many horses which supplied the power for farming until the 1920s.*

*Since I was the eldest in the family, it became my job to help Dad by driving an outfit in the field. What I remember best, was driving a binder pulled by four horses in the two fall seasons when I was 15 and 16. [That would be in the mid-twenties, one being the year John visited Norway.]*

*Once, Bernhard and I were left to cut a field of flax, when Dad was out threshing and we didn't have the tamest horses either, for those had to go on the threshing team.*

---

32   Within each township, a specified amount of land was designated for the Hudson's Bay Company, as well as for school land and for railroad.

*Two horse outfits ready to work.*

Additional details about this same incident were recorded in the personal story my uncle Bernhard shared in the community history: [33]*I have early farming memories ... of driving a binder in 1925 or 1926, when my sister drove one, too, of having lots of runaways with horses- a rake, a disc and a hay rake, twice in one day; and in haying - there were lots of flying ants when we were tramping the hay load, and of carrying coffee in a quart jar to the men in a heavy sack.*

I also remembered walking across the field to bring my dad an afternoon lunch break of a sandwich and a piece of cake, with the narrow pint jar of coffee kept warm by inserting it into one of my Dad's heavy woolen socks.

Why would they be tramping the hay load, I wondered? Nowadays, hay is rolled tightly into bales, but back then, after the grass was cut and dried out, it was stacked loosely into small piles with the hay rake, which were then pitched by hand with a hayfork onto the hayrack, so it would be very loose. Tramping hay down, by foot, would pack it together, so as to get more on the load each time, and require fewer trips to bring in the hay harvest.

---

33   Bernhard R. Hansen's story from **The Saga of Souris Valley** p 108

*As we started farming,* [John speaking here] *we were lucky to have binders and used two most of the time, so we could cut grain quickly. I bought a used threshing machine, and since I've never been a machinist, I had to hire others, and we often stopped many times a day.* (Larry Hansen said that often Jesse and Ray Erickson[34] would come and run them as JB wasn't very mechanically inclined and Ray had training as a steam engineer.)

*When the boys grew older, they became real expert at getting the old machinery to run well.*

The grain, wheat or oats, was cut and tied into sheaves, which were then arranged in a stook, to keep the grain off the ground to dry, until the sheaves are collected on a hayrack for threshing. Remember, throwing sheaves of grain into the threshing machine had been John's first job at his friend's, the Brekke place.

*Clara, on left, Edith, on right, and a friend, Alice Lokken (centre) pose behind several stooks of wheat, early 30s*

---

34    Email reply from Larry Hansen Nov 7, 2019. Ray Erickson was Larry's maternal grandfather and Jesse, his great uncle.

*Two threshing scenes west of the farmyard, 1926. The top picture is a 15-30 McCormick-Deering tractor. You can see gas barrels in the wagon behind to keep it fueled up. Larry Hansen said it is rare to see two wagons pulled up to a threshing machine like that. It had to be a really good, quiet team of horses to get that close to the belt when it was running. On the bottom he has the Avery tractor which has taken the place of the steam engine that he had borrowed in the past.*

\*\*\*

Another story, my mother, Edith, told happened about the same year. Her first outside job, when she was 14 or 15, had been when she helped with the food preparation for the harvest crew at their neighbours to the west, Bernhard and Bina Johnson. When the meal was served, Bernhard had offered wine or some other alcoholic beverage, which JB declined, but my mother observed, the preserved fruit served for dessert was definitely fermented.

*\*\*\**

JB continued- *When I built up the farm, I borrowed money off and on and didn't find bankers that good to deal with, so I was glad I got it paid up. Our Liberal government here in Canada helps the banks get rich at the expense of poor people.*

I was surprised, John didn't mention anything in his memoir about the enthusiastic support he gave to the Cooperative Commonwealth Federation, also called the CCF Party, under the leadership of TC Douglas. One of my early memories of being together with my grandfather was at Foster's Grove when TC Douglas, who also was the MLA, representing our area in the Saskatchewan provincial government, was speaking. And though I was about 5, I remembered TC's speaking style demanded attention.

*\*\*\**

Tragedy is about to unfold, leading to John's pivotal life experience.

*\*\*\**

CHAPTER 8

# SETTING THE COURSE

*A family tragedy*
*1922*

*In the fall of 1922, when threshing was over at home, I got seriously ill with typhoid fever[35]. We had been threshing for others using a cook car, a little house on wheels, to cook meals in. Before I got sick, I'd been in town and bought peaches. Likely from eating too much, three of our girls became very sick.*

*The doctor came several times and we even got Nurse Pringle, but Evelyn Violet born May 21, 1921, got brain fever and died October 22, 1922.* [Brain fever is spinal meningitis, likely a side effect of typhoid]

[Note: Typhoid, caused by a salmonella typhus type bacterium, is spread by means of poor sanitation, so it couldn't be from peaches, except if they had been contaminated and not washed properly. I guessed eating meals prepared in a cook car meant the food would be more liable to have been contaminated, due to little or no hand washing capabilities. That had been my original explanation, but the answer came from the Community History book, see previous footnote]

---

35    **The Saga of Souris Valley RM No. 7** page 695 The Preddy Family story (from Tribune). Fred Preddy mentions that he and six people altogether were sick with typhoid fever after drinking contaminated well water, during the harvest season of 1922. The Preddy family ran the Grocery store in Tribune.

*Pastor Brandser came and held the* [funeral] *service where I could hear him from my bed, where I lay sick.* [The living room was next to their bedroom.] *Yes, little Evelyn was blessed to have gone to heaven. She was buried and I wasn't able to be there. But we'll meet again in the mansions Jesus has prepared for us.*

*While I lay in bed, I saw a vision which I'll never forget. It was hell-like and I got scared and turned quickly and saw Jesus' face on the curtain. This taught me to look up to God, from whom all help comes.*

Another side effect of Typhoid was neuro-psychotic symptoms. But God can use those, too.

\*\*\*

[Later] *when Pastor Falkeid had several days of services and spoke of sin and judgement, I wanted to become a Christian, but couldn't move till young Hans Rosvold gave me a hard poke, and I fell on my knees, along with many others and accepted Christ, as Saviour and Lord.*

\*\*\*

Now recorded in the minutes of the Hauge Lutheran Church, this would have been the next year after the change in John's life? The pastor had been very busy serving his multi-point parish, as you can see from his report.

*Pastor's report from, November 1923 to November 1924: He had 139 services, meeting with the children 61 times, attended 57 Ladies Aids, 18 sick bed visitations, conducted 8 funerals, baptized 18 children. Last winter, the call* [this means the area that this pastor was serving] *had 2 teachers, Syver Oen from Minot, N.D. and Juel Marken[36] from Pennant, Sask. They had 117 days of*

---

36   My cousin, Wayne Hagen married Dolores Marken, this was a family connection to my grandfather's friend, Juel Marken. Juel was the brother of Dolores's Grandfather, Erick Marken, who with his brother Erick, left the Valdres Valley in Norway to come to America in 1905. They went to Sheyenne, North Dakota where some sisters were already living. In the spring of 1910, Erick and his brother Ted left for Saskatchewan and homesteaded near to where the town of Pennant would eventually be established. Juel came

*school for 106 children attending.*[37] *We also had last winter, special meetings and they held 19 services and made several house calls. Broder Skjervem from Carlyle, Sask* (this spelling was a guess as the writing was unclear) *was with us for a couple weeks* [as a special speaker] *and had 5 services in the call.*

*Oen, Marken and Melby have together held 18 services in the summer in southern Saskatchewan. Laymen's group held 18 days meeting in the call.* [I think this is referring to the Hauge Innermission Society.]

***

*I have in all used 525 gallons of gas - 15 miles per gallon- which becomes 7875 miles......now we go back to last year and we also this year have won several souls. God's word was spread to both adults and children. Thanks, and praise and honour goes to our God and Father in Christ Jesus and the Holy Spirit.*

At first, I thought it sounded like J.B. was also the chauffeur, but he was just the secretary, recording the minutes. It was the Pastor's report, a follow-up to this earlier item, recorded in the Hauge congregation on Mar 3, 1922: *"The "Call Committee" meeting was held at St. John's* (near Beaubier, now Melby's home congregation) *when it was decided to call O.L. Falkeid as pastor and that the congregation was going to pay for a new Ford car for him to be used in his duties as pastor. Hjalmer Johnson and D.L. Melby were elected to buy the car and H. Johnson and Ole Torkelson were elected to be on the "House Committee"* to provide for a house for the pastor to live in, also called a parsonage. A 'Call" was an official letter from the congregation making a request that the pastor selected would consider leaving where he was to come to serve in their area.

---

to Estevan around 1925, where there were more opportunities for carpentry. When the Hansen children attended school in Estevan, they stayed with Juel Marken.

37    *In the twenties, there was mention of several years when there were many special meetings as well as Parochial schools for the children.* [This would be something like Vacation Bible Schools that are held today.] Ken Torkelson shared this in the letter, referred to earlier.] Some of those meetings would have been the Confirmation classes my mother and father attended before their Confirmation in 1925.

In the next year, on page 99 and 100 of the record, they read out the names of all those that had paid for the automobile at the Dec 3, 1924, meeting at the O. Lokken place.

This next special celebratory event was the result of that motion. [38] *"Back in the pioneer days, about 1912-13 the people formed a St. John's Lutheran congregation, and a cemetery was established, a mile north of the Melby homestead. A place of worship was also built on the cemetery grounds. This land was donated by John Claybo. The church consisted of a basement with a roof on it* [and it was used] *for church, Sunday school, Confirmation classes, funerals, and evangelistic services. The plans were to eventually build a church, --lack of funds, (I suppose).*

*One time, on the 17th of May celebration* (May 17 is the national holiday for Norway) *they gave a brand-new Model "T" Ford car to the Pastor, who was originally from Norway. The car cost about $700. People came from Montana and the parish. At that time the* [Lake Alma] *parish consisted of seven congregations, covering the area from what is now including Oungre area (St. Olaf), Ratcliffe (Hauge), Beaubier (Norge, St. John), Lake Alma (Saran, Overland), Gladmar and Maxim (Immanuel). This* [event] *took place in 1923 or 1924. Mother estimated that there were 400 people there. There was not enough food to give everyone a dinner.*

*Falkeid served here from 1922 to about 1925.* (Pastor Falkeid had also served in this area briefly in the teen years. He had also been at Dell/Orrock in south central Minnesota, where the Melby's first lived after their marriage in 1901. In a 1926 clipping from my Melby grandparent's 25th wedding anniversary celebration, reference was made to them having known each other before.)

"When I was a child, we had school holidays in the wintertime," Edith described her experience. "Pastor Falkeid was instrumental in obtaining the services of two devoted Christian teachers to come to his eight-point parish to teach Norwegian Religious School /Parochial School for a couple of weeks in each place. We had Syver Oen, who came from near Minot, ND- the other was Juel Marken, who helped build the Lutheran church here on 8th Street [in Estevan.] Syver was our teacher at least two

---

38    Ken Torkelson letter

winters- teaching us the basics of the Christian faith, (Apostle's Creed, the Ten Commandments, The Lord's Prayer) Bible History, Catechism and many Norwegian Hymns and songs."

*12 confirmands from three different congregations (Hauge, Zion, and St. John) were confirmed in a Confirmation Service April 19, 1925, at Dravland School: l-r: Elmer Hillstead, Orville Fossum, Pastor Falkeid, Walter Hanson, and Victor Melby. Girls l-r: Edith Hansen, Agnes Torkelson, Pauline Saxhaug, Alice Lokken, Alma Domstad, Gladys Parnas, Alice Johnson, Helen Saxhaug.*

At the beginning, the Congregation had been served by a pastor from the east, but when the Lake Alma Parish was organized, the Zion group, south of Hoffer and the Hauge group from Ratcliffe area, amalgamated, to become the St. Olaf Congregation, eventually meeting in Oungre. The church building was the original Lyndale School building, starting in the late 1940s. The building was shared with the United Church and was also used for the Community Sunday School. At first it was located on Oungre's Main Street but was moved to the street on the south side. St

Olaf Congregation was disbanded in January of 1967. Members could opt to go east to Bromhead or west to Beaubier.

***

Faith was important to John, as was learning. His memoir briefly described his involvement in education at his home and in the community but now we will elaborate further on the Dravland School District's history.

But first, a story my mother told, fits right here. After JB's conversion, he instituted daily family devotions with Bible reading for the family as they sat down around the big oval dining room table for breakfast. My mother said she never cared for cooked cereal very much, because having been served up before they sat down, the porridge cooled off too much, by the time she got to eat it.

***

## Chapter 9
# DRAVLAND SCHOOL

Little country school houses dotting the prairie landscape were a big part of family and community life in the time period of this homesteading story. The 1915 map, introduced in Chapter 4, showed how the Souris Valley No 7 municipality was divided up into School Districts- each with their own one room schoolhouse and distinctive name. John's family was a part of Dravland's story.

John wrote, *We were very fortunate that the highway was built by our farm and the school building was only a half mile away, so our children seldom missed school.* This road had been marked out from the early years. The Dravland School district number, '947' on the RM of Souris Valley map marks the line of the east-west Highway 18 and the school location. [Highway #18 was 10 miles north of the Canada/US border.] And the dot on the eastern half of section 23 marks the location of the Hansen buildings.

*I was so glad all our children liked to learn, but when we had to send them away from home when they had gone* [as far as they could] *through our home school, it was expensive.* [Country schools went up to Grade Eight, after that, the option other than going away for school, was to take Government Correspondence School to continue and or complete their high school.]

*I was on the school board a lot and was also Secretary.*

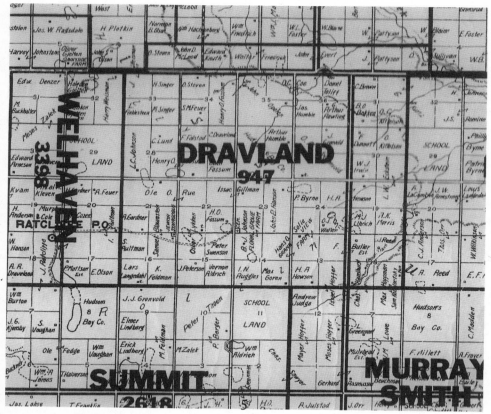

*The Dravland district from Souris Valley Municipality No 7. The location of the hamlet of Hoffer was the northwest corner of the Murray Smith District on land that had been part of the Charles Hazelhurst's homestead.*

*It came in handy that we had the telephone when the school started. We have had it many years. I was on the telephone committee. Hard times came; the telephone system fell apart, but in 1957, they were hoping to get it going again.* John reported in his memoir. So that explained how we finally got a telephone on our farm when I was in Grade Nine.

\*\*\*

In 1989, and then again in 1999, former Dravland students, teachers and their families got together for a reunion at Oungre Regional Park (or what old-timers referred to as Foster's Grove, named for the farmer[39] who had homesteaded here. Foster had planted a lot of trees down by the bridge over Long Creek, where he hoped to eventually, but never did, build a new house. After his death in 1926, the grounds were turned into a ballpark and picnic area.). Today, Oungre Regional Park is the hub of the community, a busy sports and social centre in all seasons of the year, thanks to Foster's grove of trees and a continuing sense of civic pride.

A small gray booklet, put together for that reunion, outlining the history of Dravland school district, was the source of the material I used for this chapter. I suspected, based on the writing, that Kathryn Durst Groshong, a former Dravland teacher had written it. Kathryn confirmed, she had, and that a committee had worked to put it together.[40] The map included with the booklet, showed who lived where within the district.

Evelyn Hansen, Christine Fossum Memory, and Sophie Frohlich Feldman, at the welcome desk, the Dravland booklet sold for $6.

Teachers from Dravland's History, in front of an old school banner: standing- Edith Hansen Melby, Evelyn Hansen, Bill Durick, Kathryn Groshong, seated- Mike Rooney, and Pearl Dunbar.

---

39   The Saga of Souris Valley RM No.7 Page 420 the story of Frank Foster

40   I would offer a piece of advice to individuals responsible for putting together booklets of historical information, that authors and sources be recorded.

*Mike Rooney taught from 1921-1923 and had the four oldest Hansen children as students and possibly Anna who was born in 1917, would have started school in 1923.*

*1989 reunion group.*

*Edith Hansen's class at Dravland School, from 1934.*
*1st row on step: l–r: Sam Singer, Charlie Frohlich, Ethel Singer, Anita Feuer, Rosie Singer, Edith Pell, Sophie Frohlich, Noreen Rogers. 2nd Row: Mandel Frohlich, Lionel Feuer, Lloyd Hansen, Lillian Johnson, Goldie Singer, Vivian Fossum, Evelyn Hansen, Gertrude Fossum, Clifford Johnson, Earl Hewson, George Keives. Back row: Johnny Hansen, Sam Frohlich, Leona Elfenbein, Edith Hansen (teacher), Sophie Keives, Levine Johnson, Edwin Fossum.*

*Edith, in the centre, surrounded by her former students 55 years later, at 1989 reunion.*

ELAINE MELBY AYRE

# The Dravland School District

*Dravland School District organized and built in 1912, derived its name from Fred Dravland, one of the original school board members, who lived where Ed Fossum lived. School started in March 1913 with a large class, mostly unable to read, write or speak English. Many students were Jewish children, then as through the years, until it closed in 1950.*

My mother, Edith's personal recollection pointed to her experience, with this underhanded boast, "When I began school in Dravland in 1916, I could speak only Norwegian, but quickly overcame that problem. By the time the younger ones in the family arrived, English was used at home, so they hadn't my problem, but cannot now claim two spoken languages."

The 1916 census, referred to earlier, revealed the languages used in the neighbouring homes were Yiddish, Norwegian, Russian, and English[41]

\*\*\*

And back to the Dravland School story. *Because of a lack of school records, there may be some discrepancies in this report. Arthur Humble was the first Chairman, Max Feuer, the secretary for some time, but there are two people, that come to mind, who served on the board most of the years they lived here – Max Frohlich, as secretary, and JB Hansen, as chairman.*

*For the first 18 years, the school was heated by a large Waterman Waterbury stove, in the southwest corner of the room. The boy's and girl's toilets, coal shed, and barn were out behind the school. Around 1930, a basement was made with furnace installed. An improvement, for now, instead of a trip to the little outhouse, out behind in -30° weather, there was an inside toilet.*

*A four room teacherage was built in the early 30s, so the problem of finding a boarding place for the teacher, was a thing of the past.*

*Dravland could also boast they had a well and pump, a special memory for many.*

---

41  Year: *1916*; Census Place: *Saskatchewan, Weyburn, 02*; Roll: *T-21946*; Page: *18*; Family No: *216*

\*\*\*

I wondered about pictures found in my Aunt Clara's photo album where I discovered the Evelyn Erickson, who later became my aunt had lived at Dravland [42] along with her sisters, Inez and Norma, for "they had been hired to paint the school, teacherage[43] and the barn.... the last paint job ever done on the buildings."

Johnny and Evelyn were married in California November 1950 and this teacherage was the first place they lived when they returned. I remember spending a day there the winter of 1952 after Larry was born. Perhaps this was my first introduction to babysitting the almost two dozen Hansen cousins I would eventually have and to help out when Evelyn hosted her Beaubier Pentecostal Ladies Bible study group.

\*\*\*

*In the early years, the school was used as a meeting place for the Grain Growers Association meetings, programs and picnics. Joe Ratcliff, Fred Durst and Mrs. Wilkinson were local entertainers who presented concerts at Dravland School. It also served as a place of worship for both the United and Lutheran churches and it was also used by the Jewish community for Hebrew school, taught from 4 to 6pm, after regular classes were over.* [The Jewish Synagogue for the Sonnenfeld Colony was built about 1912, several miles south, on the next road west of Dravland School, about a mile west of the Dravland Cemetery. It was moved into Hoffer around 1950.]

*Sometime after the larger school unit came into being in 1949, the remaining students in the district went by bus to Lyndale School in Oungre.* However, in September 1949, when I started Grade 1, several of them were in Hoffer School.

\*\*\*

---

| | |
|---|---|
| 42 | **The Saga of Souris Valley No.7** page 110 from Johnny Hansen's story |
| 43 | A small house for the teacher to live in. |

The beginning of the consolidation of the small school districts was marked when Dravland School and Range View School buildings were moved to Oungre in 1954. My brothers and I, who usually had walked the mile and a half to school in Hoffer, now went by school bus to Lyndale School in Oungre. It didn't cut down that much on our walking, for we still had to walk out to the highway in the morning. On the trip home, we opted to get off on the grid road a mile east of our farm and hike cross-country and chase our cattle home. Once harvest was complete our cows were out on open range pasture. That cross-country trek was better than waiting to be the last to be dropped off. So that's how several years of my school life were spent in Dravland School.

After the new Lyndale School was built in 1957, Dravland was used as a bus garage. On an April 2017 trip through Oungre, I noticed Dravland now stands open to the ravages of prairie wind and weather.

***

*Some early remembrances,* from the Dravland story, *were the school gardens in the early 20s, the produce was entered at the school fair held in Tribune in the fall. Other school projects such as writing, map drawing, etc., also brought coveted ribbons.* (One of the projects, among my mother's memorabilia, was a handmade booklet, listing names of various wildflowers with dates and locations of sightings. Finding this brought me an extra special feeling of kinship with my mother, since searching for wildflowers throughout the season was a favourite activity of mine.)

*The grade eight pupils had to go to Tribune to write their final exams in the 20s.*

It was an adventure for Edith, who shared this remembrance in one of her history bites, "When I finished Grade VIII in 1924, I had to go to Tribune to write my exams. Long Creek, which we had to cross below the Geo. Kurtz place, was in flood with deep water over the bridge. We drove across. What a chance to take!"

And about ten years later, the experience of their neighbour's daughter was memorable in a different way. To get the money to pay her fees for the exams, now being written in Oungre, "Rebecca Frohlich took in a pail of gopher tails to the RM office. In an effort to cut down the rapidly rising gopher population, devouring the dwindling vegetation due to the drought, the R.M. paid ½c per gopher killed, as evidenced by supplying the tail."[44]

There was even a competition between the schools, as to who got the most gopher tails, according to the history of the Shakespeare School district with the prize awarded at the community School picnic.[45]

Another similar natural phenomenon, from this very same time period deserves mention: the 'Rabbit Drive of February 10, 1934'.[46] I'm sure it was a topic of conversation around the community and among Edith's students for it took place one of the years she taught at Dravland. Rabbits were overrunning the country. A pen was set up at a central location and numbers of people from the community walked or rode horses in a line cross country, heading towards the closed in area where, on this occasion, 2800 rabbits were trapped. With the shortages of so many things in the thirties, I should imagine that there was likely a lot of rabbit stew. I read someplace, but now can't find the reference, that money from selling the fur pelts after the rabbit drive was used to buy seats for the Community Hall. These drives happened in many other areas as well.

***

My mother, Edith often recalled a special school trip to Regina, on King George and Queen Elizabeth's 1939 Royal Tour of Canada, after their coronation.

Edith would have been at another school then, so I asked Chuck Frohlich, "Did Dravland make such a trip?" And in a Dec 26, 2017 email reply he said, *"Yes, our school went by truck to see the King and Queen in 1939......we*

---

44    The Saga of Souris Valley No. 7 page 103 from Max Frohlich family story.
45    The Saga of Souris Valley No, 7 Shakespeare School District No 3057 by Henry Good p 613
46    The Saga Of Souris Valley No. 7 Picture on page 758

*stayed overnight in a big building (might have previously been a barn) and slept on the floor. I can't remember if we slept on a blanket or something else. Most of us had never been that far from home before. (We didn't have a car yet, so the farthest we would have been from home, would have been a radius of 4 miles or so)."*

\*\*\*

*In the winter of 1947, an airplane crashed a distance south of the road. The roads were bad due to the snow. 12 airmen stationed in Estevan, slept at Dravland School and had meals at the Hansen's while waiting for the many days to get to the crash site.*

\*\*\*

*No one can forget the excitement of the annual school picnic the Souris Valley Municipality held in the spring at Foster's Grove. Both as a student and as a teacher, I remember the buildup: practicing the marching drills and the excitement that increased until that final glorious day at Foster's Grove.*

*Thrilling band music gave our feet a lift as we marched, with each school holding their banner high. We took part in the individual sport events with determined effort, straining every muscle to win ribbons and medals.*

Highlighting this last statement was another page of Edith's notebook that showed the interschool competition was based on anywhere from 7-11 items. Marching was the highlight. This was most likely from when Edith was at Dravland, for it is listed first. The total score was divided by the number of students- Dravland came out fairly high at 3.9, but Greenmount with 4.9, and Shakespeare with 4.6, took first and second place for the highest overall score.

*For one splendid day, the drought, grasshoppers and dust storms were forgotten, it was a bright and happy day, each year in our lives.*

\*\*\*

The legendary Dr. Brown, along with his wife, Laura, dedicated their lives to the welfare of the residents in and around this broad area. He was credited with the idea of instituting the Souris Valley School Picnic, which motivated people from all corners of the municipality to work in combined effort for this singular event. In the depths of the depression, when folks did not have the wherewithal to pay for the license to put on their cars or trucks, Dr. Brown arranged for a special dispensation with the authorities, so everyone got to come out and enjoy the picnic.[47]

And as we continue, we will look at the further education and training that was part of the family member's experience.

---

47   The Saga of Souris Valley No. 7 page 7-11

CHAPTER 10

# BEYOND THE LITTLE RED SCHOOLHOUSE

*1926 and on*

"Since there was no opportunity nearby for high school," Edith wrote, "I took my Grade IX, X and until Christmas of Grade XI at Outlook Lutheran College. I finished my Grade XI at Estevan Collegiate in 1927. Myrtle, who was one year younger, was also there with her the years my mother was in Grade X and XI.

Looking out the southwest window of her Trinity Tower apartment in Estevan, my mother, Edith explained as she pointed to the house on Second Street where the family lived, those several years they attended school in Estevan. "It was second from the corner" (across from Trinity Lutheran Church) "but had been the identical construction to the house [now] on the corner."

Both Edith and Myrtle had been students at Outlook College. The time my mother spent there had been very special to her, so I realized how disappointing it had to have been for her to have to leave partway through the term. My aunt Clara also told me how unhappy she had been to leave all her friends in Dravland behind, to attend school in Estevan. The elementary school was just down the block and across the street but the Collegiate was at least 16 blocks west.

Having to go to school in Estevan was only a stop gap measure. Within a year, there was a high school classroom in Hoffer where first Myrtle, Bernhard and Clarence and, then Anna and Clara took the balance of their high school and Johnny took part. However, the high school room in Hoffer closed in 1939, so Lloyd took his high school through Correspondence School, then for the first part of Grade 12, he rode his bike the 4 ½ miles into Lyndale School in Oungre, but when Edith and Victor came home for Christmas in 1943, Lloyd returned with them to finish off the balance of the 43/44 school year at the school in Birch Hills where they now lived. (And when I started school at Hoffer we were glad to have that extra room on the west side to play games like Red Light-Green Light, Giant Steps or Mother, May I?)

Palmer, the youngest of the Hansen family, completed high school at SLBI in Outlook, SK; then competed a Bachelor of Commerce degree at the University of Saskatchewan and became a chartered accountant in Saskatoon for all his working life.

Evelyn attended high school in Lake Alma up to Grade Eleven, then took Normal School during the summer months in Regina. She got her teaching certificate after 2 months of teacher's training and started teaching at the age of 17 at Prairie Lea School, near Creelman at the end of August that year (42?). Eventually, she did complete her matriculation, which she needed to complete her Bachelor of Education degree. She and Myrtle located my boarding place for my second and third years at the College of Home Economics, University of Saskatchewan in Saskatoon for I had started the first year of my degree at Luther College in Regina. They were completing their Bachelor of Education degrees in the early 1960s, as I started mine.

Bernhard took a year of junior college at Outlook College in the late thirties. By then, the name had changed to Saskatchewan Lutheran Bible Institute (SLBI) and later became Lutheran Collegiate Bible Institute (LCBI) by the time I attended there in the early sixties.

So, of the family of ten, the only ones short of a grade twelve diploma, were my mother, Edith, and my uncle, Johnny who only got their Grade Eleven.

## Edith's story

Another of my mother, Edith's little 'History Bites" fits in here: "*The next year, 1928, I attended Normal[48] School in Regina, after which I taught school for 10 years [at] Round Grove, Velhaven, Dravland, Shakespeare, Blooming, Workman and Greendale - from 1929-1940 with one year off- six months of which, I spent in Seattle with my uncle and aunt, Jens and Mathilda Gronvold, and three months doing housework, at Vancouver, BC.*

*I started teaching with a salary of $1000 a year, which dropped to $500 in two years, but it had raised to $750 the last year. Much of the wages we had to wait for, until the Government grant came, sometimes many months after we had earned it. Room and board was $15 a month. Even that, couldn't always get paid on time. I had a note from one school which was paid up after I was married- a wait of ten years.*" A page from Edith's mini record book revealed these 'notes' averaged $12.95 a month, when she was at Shakespeare school.

*\*\*\**

In 1998, as I researched to write the Melby family history for **Settlers of the Hills,** the Lake Alma RM No. 8 history book, I was in contact with Ken Torkelson[49] who spoke about my mother, Edith, as the best teacher. She had been his teacher at Velhaven School for the two years when he was in grade seven and eight. One time, she was standing by his desk, explaining a concept to him, when suddenly, she fainted, and fell on his desk. He was shocked and concerned. She came to and went over to the teacher's desk, where she sat with her head on her arms for the rest of that school day.

Ken said, "At Velhaven, there were 13 children from Norwegian families, and 13 from Jewish families, who had come over from

---

48  "Normal" was , strange name for the school that trained teachers.
49  Letter from Ken Torkelson, referred to earlier

Europe in the 1920's. In the good weather, Edith drove a buggy back and forth, and the students would unhitch the horse for her, and put it away in the barn. During the winter, she stayed with his parents, the Ole Torkelson family."

Edith always hoped to be able to teach out of the home district, but the farthest she ever was away from home, was when she taught at Workman School at Carievale, where the Hansen family friends, the Leonard Johnsons were living at that time. Her last school was Greendale.

***

Three other family members made significant contributions to education in the province. Myrtle taught at many country schools starting about 1931, including Hayland School at Bromhead (her first teaching job), Maple View School (Torquay), Diamond Coulee School (Minton), Hamar School (Bromhead), and Crooked Creek School (Arcola) and Roseville School (Claybark).

On a 2016 visit to Camrose, I met Sylvia Pederson Espe and we exchanged copies of our books, each about our own individual family history. I learned Myrtle had been Sylvia Pederson's first grade teacher. Her parent's home, the Arnold Pederson's, near Torquay, was where the teacher usually had board and room, but Myrtle chose to stay in the teacherage, attached to the school. So, young Sylvia was allowed to stay with Myrtle there, so she would have company.

Later, Myrtle taught high school in Lake Alma for 2 years and Frontier for one year and in the late 40s, Myrtle was asked to come and teach at the Lutheran Collegiate Bible Institute in Outlook, Saskatchewan, where she spent 23 years, retiring in June of 1976.

Myrtle had been my English literature teacher at LCBI, when I took my Grade Twelve year there in 1960-61. Though I knew students' name for her was "old stone face," everyone had the utmost respect for her ability as a teacher. She could get the lesson across. What a formidable teacher she was! I heard her explain once, that her 'gruff persona' was her way of coping with life as a single woman.

\*\*\*

Bernhard went to Normal School in 1939 after his two-year interval at Fort San, and taught at Apollo School, near Maxim as well as at Hazenmore. He went to University of Saskatchewan from 1942-1945, and following that, took his degree as a naturopathic doctor from National College in Chicago. He practiced in Vancouver and New Westminster, BC for a time before coming back to the farm. In 1952, he was principal at a school in Duff, SK. After coming back to farming fulltime, he spent ten years as a Radville School Unit Trustee from 1957-1968 during the time many of his Hansen nieces and nephews attended schools in the area.

Bernhard started spending his winters in Arizona, where he met and married Hazel Scheske on Feb 14, 1981. They continued to spend summers in Saskatchewan, first at the former Ed Block place and then in Weyburn and the winters in Arizona until Bernhard passed away April 30, 2002.

\*\*\*

My aunt, Evelyn Hansen was the last teacher at Dravland School. I was probably five, when she brought me to school with her one day- the limit of any Kindergarten for me. At that time, only half a dozen students attended there.

Mazes of tunnels, made by those students in the huge snowbanks in the trees surrounding the schoolyard, created quite an impression on me, something I aspired to construct as a child, but never managed to recreate.

And then one of Edith's handwritten history bites helped me identify the time and circumstances.

*In the winter of 1949, I* [Edith] *had come home from the hospital after a varicose vein operation, when Vic had to be taken to hospital the next morning for an appendectomy.* [My uncles] *Bernhard and Clarence took him by way of*

*Weyburn, for Highway 18* [to Estevan] *was blocked. High walls of snow on either side of the road made driving a hazard.*

So that initial school experience must have been this snowy winter of 1949, as by September of that year, I had started Grade One in Hoffer. And the Melby children were likely being babysat at the Hansen farm. Amazing, what I learned, putting this story together!

\*\*\*

Clara and Anna attended SLBI at the same time in the middle forties. Anna graduated from the two-year Bible school program in 1947[50] and worked as a parish worker and a dressmaker before her July 1952 marriage to Olaf Hagen.

My uncle, Harald Melby had been pastor at the Lutheran Church in Swift Current, SK the year Anna was a Parish worker there and he told me that Anna had conducted a rather significant survey of the members of that congregation.

Sometime, in the summer before she was married, my aunt Anna babysat the Melby children, when my parents traveled up north to a wedding on the Melby side. Two memories stand out. We all had chicken pox. Since we weren't feeling that sick, we tried to go over to the neighbours to play and they told us, "Go home!" The other was that Olaf Hagen came to visit Anna during the time they were courting. We were all packed into the back seat to go for a Sunday drive and teasing comments from that Peanut Gallery in the back seat were ignored!

The next year, Anna married Olaf Hagen in a double wedding along with Olaf's twin brother, Herbert and Emma Wee on July 12, 1952. Anna and Olaf farmed northwest of Lake Alma.

\*\*\*

---

50   Wayne Hagen's wife Dolores's mother was Juverna Olson Marken, Anna's best friend at SLBI. She graduated with her but missed the ceremony because she was registered to begin Nurse's Training which started on Graduation Day.

Clara had taken a secretarial course in Regina and after that year of Bible School at SLBI, she worked at the Lutheran Seminary in Saskatoon beginning September 1947 until just before her marriage to Harald Tangjerd in July 1949.

A typed postcard dated August 20, and posted August 22, 1947 from Clara to Mr. and Mrs. V. R. Melby at Hoffer, Sask, stated she had been staying with some of my Melby relatives at Kinistino, who took her to catch the train to Saskatoon. She was to take the place of Pastor Dale's[51] secretary, who was leaving her position, Sept 1. A side note said she was heading for work at the Seminary just after posting this.

***

These education matters brought us twenty years ahead of the story, so now we will backtrack in John's memoir where he describes his trip to Norway, twenty-two years after leaving.

***

---

51      Pastor Mars Dale was the President of the Lutheran Seminary in Saskatoon, SK.

## Chapter 11
# GOING HOME

### *1925*

John's memoir included the adventures he encountered on his trip back to Norway to visit his homeland.

*After I became a Christian, I began thinking of going home to see my dear mother who had suffered much because of my sin. Pastor Falkeid urged me to go. When the day of departure arrived, May 18, 1925, a neighbour, Oscar Lokken took me to Estevan. From there, I took the train to Winnipeg where I bought my passport and looked over the city, in the one day I was there. That was a much simpler process than the two times I have had to get a passport.*

\*\*\*

*Then, onto New York, where I was to meet Ole Heskin, in Norway House in Brooklyn. I asked a worker at the station if he'd phone Norway House. But, when he said Norway House was filled up, he said he'd find me a place. I took his offer. He asked if I'd like to see the city, the next morning.*

*He called me in the morning, and we were off, suitcases and all. I became a bit suspicious of them, when no one spoke Norwegian, where we went to exchange the Canadian money for kroner. When he crossed Brooklyn Bridge, they began to turn to the right and when there appeared to be no houses, it looked like they were going towards the ocean. I sat up and asked sharply, 'Where are you taking me?'*

*Finally, they turned left. I asked, 'How much do I owe you?'*

*He said, 'Twelve Dollars.' And the other fellow also wanted twelve dollars.*

*I refused, 'You've driven me around all morning, and I haven't even had anything to eat.'*

*When they stopped at a light, I quickly got out. When I told others what happened, they said, 'you sure are lucky to come out of it that good.'*

*In the office, I found out how I could get to the railway under the ocean* [he likely means the New York Subway] *and finally came to Norway House, where I met Heskin.*

\*\*\*

*Later, I was told I'd been cheated on the money exchange. A chap, named Nels, went with me, and I got back all, but five dollars. So, I was out $17 because I trusted a man, but I thank God for the experience. Our wonderful God watches over us and wants us to always turn to him. Yes, there is a continual battle between good and evil. The devil has many helpers but it's oneself that is most harmful, as I discovered the first night on board.*

*Someone told me, 'You are crazy leaving your wife and seven small children. You will likely never return alive.'* [Dina was pregnant with Evelyn at the time. She was born later in October of that year.] *And with my experience in New York, I almost changed my mind. But when I took it to God in earnest prayer, I continued to enjoy my trip in joy, as I knew I was doing the right thing.*

*Heskin, an old lay preacher and I sang nearly every day.* [I think this was three people, for he already has named Heskin, so I think, the old unnamed lay preacher was someone else he met on the boat.] *We landed at Bergen where we saw many people coming out of church and found out later it was Hallesby, who had spoken.*[52] *We heard him* [Hallesby] *later at a Bible Week in Vang. We went there on an electric train and Bergen was beautiful with long green grass.*

---

52    (Hallesby was an influential Conservative Norwegian Lutheran theologian, author and educator, who wrote 67 books but was most known for his devotional writings) https://en.m.wikipedia.org/wiki/Ole_Hallesby

*I'll never forget one place, so beautiful and green with bits of snow. There were several Swedes with us who had been in the United States. They, too, said they never seen anything so beautiful. In a short while, we were in a regular snowstorm which left snowdrifts several feet deep in June. I love Norway with its cliffs and mountains- my fatherland.*

\*\*\*

*When we got to Drammen, where as a youngster, I'd hauled milk to the Skoger dairy, many memories came to mind. I'd gone from bar to bar, looking for my father, who drink ruined, and he died at 44 years.* [John would have been in his mid-teen years then.] *'Yes, sin's punishment is death, perhaps both for body and soul. I hope I'll meet him in heaven. My uncle, Jorgen Nelson Borge had visited him several times before he died.*

Since John's father died because of falling out of a wagon, that last statement leads me to believe he survived the initial fall, but died later, as a result of his injuries.

\*\*\*

*I took a bus to Skoger. We stopped at Flaten Vestybygden School where my father and all my brothers and sisters had gone. I had gone to Folk School there, too, when Jordhoi was the teacher. When I got home, my youngest brother, Broder Jordbrek, had been teaching at Flaten for several years. My mother was living with them, and they didn't know I was coming, so you can imagine their surprise. It nearly overwhelmed my mother when I met her at the door. Yes, it was wonderful to come home, after having been in America for 23 years. It was great to be with my dear mother, all my brothers and their families also Jacob Ness, my only sister's husband and children. Anna had passed away when Astrid [her daughter] was only five.* [Astrid's daughter, Bjorg and I became pen-pals after my aunts, Myrtle and Evelyn visited Norway in 1956.]

*John and his mother, haying the old fashioned way.*

*Having our home on wide open plains, it seemed to me that the houses in Norway had been moved closer together since I left in 1903.*

*I'll never forget when my niece, Jenny Jordbrek and I raced down the snow-covered hills. I had worked for my Uncle Julius Gundeso, who set me a good example, to stick by my work. I was at my mother's relatives most of the time [on this trip.]*

*After having experienced my home on wide open plains of Saskatchewan, it seemed to me that the houses in Norway had been moved closer together since I left in 1903.*

*Adolph and his wife, Borghild with Bjarne 1917, Willy 1919, Reidun 1920, Erling 1921-1925. Adolph, who had been in Saskatchewan, returned to Norway in 1913. 1924 photo.*

***

*Since Heskin and I decided we wanted to go to Vang to hear Hallesby, I took one week, out of the six, I had in Norway, to go to Fagerness by train. I got a ride with some people who rented a car to go to Vang. Even, if it was July, there was such a raw cold wind, it went through our clothes and raincoats. So I had to buy some long un.derwear. Yes! You Norway!*

*I visited Dina's relatives at Aadelsbrug. Old Mr Ghilemoen took me to Torpestuen, my father-in-law, Johan Gronvold's birthplace. He showed me an apple tree that Johan had planted.*

*One Sunday morning, taking some folks out in a rented car, we had to swing to the side to miss driving over a man's head. He lay with his feet towards the ditch. Later, in another place, they said, a young boy had taken off all his clothes when coming from a dance. Yes, America isn't much better! I think it's high time for a world war on worldly pictures, smoking and drinking, they take so many lives and break up so many families.* Knowing JB Hansen's history with an alcoholic father, helped me understand, why he went so far in the other direction although with the way things are going in the world, he wasn't really wrong.

***

As JB concluded this section, he talks about getting acquainted with a distant cousin who also had emigrated and, like John, was paying a visit to his home country. *I was with Dick Jul on my trip home to Norway. After my Norway trip, we visited them several times at their home near Scobey, Montana. His grandmother and my mother were half-sisters.* [John's Grandmother had been a widow with *several* children when she married John's grandfather.

Check back to Chapter 1 and see the picture of her second family of 3 boys and 6 girls, her first family would have been adults on their own by that time.]

*Once, when we were at Dick Jul's place, I realized they had a machine you could sing into and play it back. I took the opportunity to sing a Christian song that they could play back.*

Most likely what he recorded was this song called, "My Mother's Prayer," for I had often heard my grandpa sing this[53] at family gatherings. It seemed like this song described his actual experience.

And Dick Jul was one of the visitors at my grandparent's place, at their 50th anniversary party in 1959.

Meanwhile, the homestead is a happening place, and we need to catch up with five more family members born in the 1920s.

---

53  Judson Wheeler Van De Venter **The Cyber Hymnal 4437** Public Domain

# My Mother's Prayer

Public Domain
Courtesy of the Cyber Hymnal™

## Chapter 12
# THEN THERE WERE TEN

### Family and Friendly Neighbours 1920-1929

*Johnny Henry was born on February 11, 1923, Evelyn Alida- October 25, 1925, Lloyd Walter- October 13, 1926 and Palmer Donald- February 21, 1929. God has blessed us with 11 children, ten of whom are still alive. They have all been good workers and liked school. I was always glad for the many things they had to do on the farm.*

*When Dina had her babies and in busy seasons, we were very fortunate to always get one of the girls from the Leonard Johnson family to help in the house.*

### Johnson Family

The Leonard Johnson family's original homestead had been two miles west of the Hansen place. They had been their neighbours from the beginning.

JB wrote about a visit to these neighbours. *I remember one March; I went to Johnson's to buy potatoes and they asked me in for dinner. In a little while, it was such a bad snowstorm. I started out but had to stop at another farm until 4 o'clock. I started out again and even though the storm was still bad, I made it. When I got home, everything was done up. Dina had been washing clothes and* [when she] *saw the storm coming, she wasn't long in getting the cows and horses in. One farmer lost several animals in that storm.*

Victoria and Mable, the two oldest Johnson girls and most likely, the first ones to be 'hired girls' for the Hansen's, reported, [54]*"It was a great thrill, to walk the two- and three-quarter miles* [east] *to the store operated by Mr. Gronvold, where we could buy groceries and other treats."*

So, it must have been doubly delightful for their family in 1924, when the Johnson family moved to the Gronvold farm, northeast of the Hansen homestead, where Dina's father, Johan Gronvold had formerly operated his store. The big house there had been built by Dina's brother, Olaf Gronvold, when he married Christine Gjelten in 1913. Then Johan and Maria moved to a new place, about 3½ miles further southwest of the JB Hansen Farm, and their country store was discontinued.

\*\*\*

*Dina's* brother, Olaf Gronvold farmed until the early 1920s, when he moved to Seattle, Washington. In the early 1950s, he came back to the farm for a short while, then the yard was purchased by Nick Keilbasa. So in my time growing up on the farm in the 1950's, we always knew it as Nick's Place though I did recall my great-uncle Olaf being around the farming community for a short while.

There were 13 children in the Leonard Johnson family: Hulda and Victoria had been born in Norway, while Viola, Mable and Clara arrived after the family moved to the United States in 1903, and the rest were born after the family came to Canada in 1909. Alice, Evelyn, Carl, Clarence, Levine, Clifford, Norma and Lillian Johnson would all have been classmates and friends of the Hansen Children. Following the tragic drowning of their

54    *The Saga of Souris Valley RM No 7*, original edition page 115

son, Clifford, in 1936, the Johnsons moved to the Careivale area of southeastern Saskatchewan in 1937. Both Carl and Clarence served in the army. Clarence Johnson was killed overseas in 1944.

***

A special memory, selected from my mother's photo album, showed the Hansen and Johnson families travelling together on a wagon trip. That looks like Bernhard holding the reins on the second wagon as they head northwest to the hills, in the Maxim area, on a berry picking expedition on the Dominion Day (now called Canada Day) holiday. Imagine my surprise, when my mother, Edith's small notebook diary confirmed this event: July 1933- *Sat 1-Went with Johnsons to pick berries. Finding none,* [probably due to the drought conditions] *we visited Christianson's.* [They were a family, recently arrived from Norway, who lived in the Velhaven area, immediately west of Dravland school district.]

They would have been looking for Saskatoons, later in the summer, it would be chokecherries. Both native berries grow along the sides of coulees in many places on the prairies.

*Going Berry Picking*

Now I'm getting ahead of myself, for there is still a lot happening in the twenties.

*The Hansen and Gronvold cousins on a visit to Plentywood. Johnny is the youngest in the family on this photo, so this is likely 1924 or early '25, perhaps just before John's Norway trip. Looks like a playhouse, they are standing beside? The Jens Gronvold children are Alvin 1912, Walter 1914, Evelyn 1918, Marian 1920 and Raymond 1922 Back l-r: Johnny, Myrtle, Alvin, Bernhard, Marian, Walter, Clarence, Edith. Front l-r: Clara, Raymond, Evelyn, Anna, and (from the Henry Gronvold family) Irene holding Martha.*

*A quick visit to the grandparents and Uncle Adolph, in the busy time about 1924 or 5. Clara is between her grandparents, Johan and Maria Gronvold. Dina is holding Johnny.*

Maria passed away August 28, 1928, and Johan, two years later, when he was down in Montana visiting his Plentywood families.

Edith remembered that as a small child she'd sit on her grandfather's knee and comb his beard. She claimed he liked that.

Uncle Adolph would ski the couple miles cross country to visit at the Hansen's in the wintertime. I remembered great uncle Adolph for his distinctive contagious laugh.

\*\*\*

*Choir Practise at Walter and Anna Hanson, 1927 from my mother, Edith's album.*

I wondered if this had been a one-time or a regular event, but since I knew Anna played the piano, it may have been a regular event. Anna Hanson was my father's oldest sister. They lived in Greendale, the farthest northwest corner of the Souris Valley Municipality #7. Anna and Walter Hanson moved north to Weldon in 1931, the same time as the Ole Saxhaug family.

Back two rows: Face hidden, __, John B. Hansen, Behind him? Beside him in profile ?, Walter Hanson, __?__, Mrs. _?__, Bernhard Hansen, _?__, Mrs.__?__, Victor Melby, Odin Torkelson, __?_, James Torkelson, Clarence Hansen, _?__ in front and Ken Torkelson at the end.

Front row of adults: Anna Melby Hanson, ____, ____, Ruth Torkelson (Ryan), Evelyn Johnson (Ball), Anna Hansen (Hagen) Children: Dorothy Hanson, Jeanette Torkelson, Donald Hanson, Phyllis Hanson, Alice Hanson, and Edith Torkelson

\*\*\*

*Special visitors*

One time, in 1928, the Hansens got some special help when they had a disc that didn't work well. Dina's cousins from the Espe family (who were part of the story in my first book) paid a visit to the Hansens. On the left is Christian Espe beside his wife Jesse and on the right is Albert Espe and his wife Mathilda and daughter Glenda. Albert was a talented machine inventor, who ran a machine manufacturing company in the Crookston, MN area. He was able to show them how it was built wrong, and he fixed it. (I thought Albert looked a lot like my great-uncle Adolph Gronvold.)

Albert's four sons-Henry, Clarence, Mylan and Roy, along with daughter, Glenda ran this Crookston machine shop well into the middle of the 1960s.

My cousin, Larry[55] said my uncle Bernhard recalled that visit very well.

---

55   From an email from Larry Hansen, November 24, 2017

*Clara posing with a 36 run Cockshutt drill, about mid-thirties.*

Some of the erosion and damage of the Dust Bowl was attributed to using this implement. For when the prairie, covered for centuries by native grasses, was broken up and pulverized, the dry and light grains of soil are picked up by the incessant winds on the plains.

\*\*\*

Some land dealings didn't work out as John reports. *In the fall of 1927, I bought the School Land[56] on auction (320 acres, and 40 acres of that was over the railroad track).*

*Then, I bought 320 [acres] further south-east which I sold. I left home at 5:00 in the morning by car, and broke [the land] with the tractor. I had my meal with me and ate outside. I did all this hard work for nothing when he didn't pay me……*

Land deals gone wrong, along with the poor growing conditions of the Dirty Thirties made farming discouraging and was a factor that forced many farmers' moves to greener pastures.

---

56　**The Saga of Souris Valley** Page 4. Note that in the early surveys, two sections in each township were designated "School Land" and the Hudson Bay Company was given about 7 quarters in each section. The Hudson Bay land was north of the Hansen's home quarter.]

And before we move on into the thirties, here is a good place to show off the younger crew of John and Dina's family.

*1929, Lloyd, Palmer (the baby), Evelyn in front, Clara, Anna in the back with Edith and Jeanette Torkelson, standing between them.*

*l-r -Clara, Anna, in the back , Evelyn, Johnny, and Lloyd 1930- on the veranda on the east side of the house*

\*\*\*

With the Crash of 29 leading the way, the third decade of the Twentieth Century, rears its ugly head, bringing troubles, but there was a surprising variety of good times mixed in.

\*\*\*

CHAPTER 13

# THE THIRTIES: MAKING GOOD WITH THE BAD

*Travelling around the province for regional and provincial Luther League activities.*
*1933*

The Thirties started out memorably as my mother, Edith described: *In 1931, while Father was taking the census,* [57] the wind *took the top off his Model T Ford. Shortly after that when there was no money for gas it became a 'Bennett Buggy,'* named for the current Canadian prime minister of the time. In the United States, a similar vehicle was called a Hoover Cart, named for their president at the time. The car engine was taken out and it was turned into a horse drawn vehicle.

The Dirty Thirties presented the usual challenges we associate with that decade along with extra trials thrown in as John reported. *In the fall of 1932, Dina had a breast removed in Minot Hospital. I stayed, till she was ready to come home, even if I could afford only one meal a day. Since we owed the doctor, I hauled in two loads of wheat at $.30 a bushel, the least I'd ever*

---

57   **The Saga of Souris Valley** page 108 from John B Hansen story

*gotten for wheat. Now we get about $1.50.* [One bushel was 60 pounds, or 27 kilograms. Remember he's writing this in the late 1950's]

*The doctor's bill was $55.00 and the hospital bill was $34.*

\*\*\*

Jacob Stolee worked at the JB Hansen farm one summer while he was a student at Lutheran Seminary in Saskatoon. In between, he helped the pastor. Judging by a picture taken with the two youngest Hansen boys playing in the barn it was in the earliest thirties.

Thirty years later when I attended LCBI for my Grade Twelve year, Rev. Jacob Stolee was the President of that church school.

The Ole Torkelson family spent time together with the JB Hansen families fairly often, judging by the number of pictures of them together and on this occasion, Jacob Stolee was there, too. L-r: Bernhard, Lloyd, Jacob, Odin Torkelson, Agnes Vinge, James Torkelson, Edith, Anna, Clara with Edith Torkelson and Evelyn in front. In the back, Ken Torkelson and Clarence.

The Years [58] of 1931, 1934 and 1937 were the no crop years while 1932, 1933, 1935 and 1936 were the poor crop years. The price of wheat went below 30 cents per bushel in 1933.

My mother, Edith's entire working career was this decade. In one 'history bite' she had written, she states as matter of fact, *"These were the depression years, and no one had any money. Relief, food and clothes were shipped in from more well-to-do areas down east, for which we were thankful. Young people didn't rush off to the cities for jobs- there were none- so we were not at a loss for friends and companionship. We made our own good times."*

*Young people having fun. Alma Nelson sitting in the middle seems to be the centre of it all. I can identify Clarence, Bernhard, Myrtle and Clara Hansen, Victor Melby, James, Ken and Odin Torkelson, Sanford Johnson, and Olaf and Herb Hagen in the plaid shirts. Someone stole Herb's hat and his farmer's tan is showing. This picture was taken by Edith.*

---

58    **The Saga of Souris Valley No. 7** page 108 from John B Hansen story

## ELAINE MELBY AYRE

***

You were introduced to Edith's mini diary beginning with Berry Picking, July 1. Then from that date she outlined the busy summer of '33. The next day, July 2, they were busy preparing for a Luther League gathering for young adults and teenagers at Viceroy, as summer holidays started, and it ended with a province wide convention in Birch Hills.

How did they get there? Viceroy, on the northern side of Willowbunch Lake, north of the Big Muddy region was about 175 km or 110 miles to the northwest of where they lived, going west on Highway 18, then connecting north with Highway 6, though I thought they may have been taken other side roads. In that time, the typical means of getting to events several hours away, the entire group travelled together in the back of a large truck. Benches provided seating and other than sideboards built up around the truck box, there was no additional protection from the elements.

I wondered how many from the family were along on this planning activity, and this notation gave the answer: *July 2 - Anna, Evelyn and I drove* (with her horse and cart) *to Harold Torkelson's to arrange for Viceroy Trip.* Since Evelyn would have only been around eight years old, she probably went along to play with the other younger girls in the Torkelson family. From that date, Edith listed events all the way to July 31, giving us a glimpse of the busy summer of '33.

*3, 4, 5- Preparing for Viceroy Trip. [Made at home]*

*6- Left for Viceroy - waiting- truck broken - rain - pitched a tent - camped two hours - took the wrong possible turn - truck loads - Camp out Surprise Valley School, slept on the floor after breaking open the basement door.*

Surprise Valley was the name of the municipality, west of the RM of Lake Alma, the neighbouring municipality west of the Souris Valley RM. It was also the name of the one room school, somewhere near Gladmar that sheltered this group of young people that rainy night until the next morning's daylight gave the driver a clearer idea of the lay of the land and the way to go.

*7- Arrival at hall. Rev. Melsaether giving orders, while we were dressing in the stuffy car. Enjoyable meetings.*

*8-9- Stayed at O. Bakke home- slept on floor.*

10- [Went] *with Vinges to John Olson's at Minton to prepare for the trip to Birch Hills.* [Rev. A.M. Vinge was the pastor serving this area, first as a student-pastor in the summer of 1930, and then full-time, from 1931-1934. The young people's group, called Luther League, was most active when Rev. Vinge was here in the parish. This event was the Luther League provincial convention. Five family members attended]

*11- Agnes Vinge* (the pastor's sister) *accompanying us to Birch Hills- leave Olson's at seven- arrive at Birch Hills at 8:30 with only a half hour stop in Regina. We go to Reverend Langley's to be billeted at Andrew Olson's, northeast of town.* Sounds like that trip took over 13 hours. I remembered this trip as taking no more than 8 hours when our family made that trip to visit my Melby aunts, uncles and cousins in the 1950's. I presume the roads and vehicles were better by then.

*12- Very warm morning but cold rain shower before evening. Met many old acquaintances.* They were classmates from when Edith and Myrtle had gone to Outlook College in 1926 and 27. Reverend Falkeid, who had served in the south 1922-1927, and had confirmed her, was now the pastor in Birch Hills. The Melby family had moved up north several years before as had other families from the south.

*Meet Helen Mickelson, Lloyd Slind, Einar Berg.*

*13- Very cool day. Reverend Aasgard, President of the Norwegian Lutheran Church of America speaks very ably and interestingly.*

[It was amusing to find this comment about Rev. Aasgard, this special guest speaker, who had the role of leading the Norwegian Lutheran church and helping immigrant Lutherans take on the 'American way of life', and that this "must have seemed like trying to herd cats."[59]]

---

59  http://metrolutheran.org/2010/11/a-forgotten-giant/

*Met Mr. and Mrs. Bergsagel* [He was the Outlook College administrator and teacher when Edith and Myrtle attended there.]

*14- WMF meeting in English church, talk by Evangelist Lokken on the "Four Branches of our Church." College quartet sings. Lloyd treated Agnes and me to supper in the basement.*

*15- Saturday. Being sick, I eat no breakfast or dinner but the supper at the Norwegian cafe!! Einer and Gladys, Lloyd and me. The boys sing "Going Home". Sad parting, eh?* [Sounded like this was an old boyfriend from her Outlook College days?]

*16- Sunday – communion service at Birch Hills.* [This marked the end of the convention.]

*Very hot, leave with Lokkens in afternoon. Yummy ice cream and strawberries, we visited Round Lake.* [The Oscar Lokken Family, more transplants from southern Saskatchewan, were Mervin, Alice and Viola. Alice was the same age as my mother, Edith; they went to Dravland School together and were in the same confirmation class. And the ice cream was homemade!

Several pictures in Edith's photo album, show this picnic event with the Lokken family described above, and both Myrtle and Clara were on them, and maybe even Bernhard. So that answered my earlier question, about who went!

*17-Alice goes to work for Ericksons, north of Brancepath*

*18-One cup berries, oh boy, more ice cream! Melvin and Viola take me to Gjesdal's in evening.* (Margaret Gjesdal had been Edith's best friend when she went to Outlook College.)

*19- Gjesdal's - a grand, up-to-date home, Building a new big barn. Milk 18 cows. Six girls go to Birch Hills for Sports -undecided about going or staying in town. Midnight lunch when Alfhild came home from the show. Many talks about the college days.*

*20- Margaret and Selmer bring me to Melby's – difficulty finding the place in the appearing wilderness.* (Harold Melby, my father Victor's youngest brother, described one of the places the Melbys had lived as remote and very rustic. This was probably it.)

*21-Walk through the woods with Harold* (Melby), *Elsie comes and we hunt for berries. Ride home -no brakes-ride down the hill – ditch – Mule team to draw us out. Arrived safe at Hagen* [the town west of Birch Hills, where Elsie and Willie lived].

Elsie, my father's younger sister, and Willie Folstad had been married in November of 1932.

*22-Saturday- Not feeling well, sleep most of the afternoon- Chautauqua in Birch Hills in evening- Negro Singers.*

*23-Sunday. Spend the day at Prince Albert. Elsie and Willie, myself and Charlie Smith to form a foursome. Supper and dinner at auto camp. Yum! Yum! Good things to eat. Visit San* (a TB Sanitarium that opened in 1930) *and penitentiary. Treated to ice cream, then cherries and more cherries.*

*24 - Elsie takes me to Mickelson's. I stay with Selma Aune, visit with her folks. Helen teaching parochial school* (like Vacation Bible School today.)

*25 - Visited at Mickelsons all day*

*26 - Elsie comes for me – go to Birch Hills to see Marionettes in the Chautauqua.*[60] *Elsie and Willy take me to Saxhaugs.* [More former southern Saskatchewan people, now in Weldon, east of Birch Hills] *Pauline and Emil and Julstads being there, I stayed at Haave's for two nights.* Pauline was same age as Edith and they had been confirmed together.

*27- Ladies Aid at Weldon, visited with Alfhild.*

*28-Walk 3 1/2 miles to Saxhaugs from Weldon. Washed some clothes. Went with Pauline and Emil to Walter Hansen's, find Walter and Donald with a crippled leg and foot, stay here overnight.*

---

60    The Chautauqua Movement sought to bring learning and culture to the small towns and villages during the late 19th and 20th centuries. There was an Elvis Presley movie about Chautauquas called **'The Trouble with Girls'**.

*29- Saturday- Walter and children take me by car to Saxhaug's.* (Walter was married to Anna Melby, Victor Melby's oldest sister. Anna and my mother corresponded even back before she and Victor were a couple.)

*30-Sunday- Dinner at Jacobson home- Dagny, Erling, Ella, Helga, etc.*

*31- Monday- 6:45 start return trip with Julstads. Treated them to lunch at Raymore- 1 ¼ hour stay at Regina. Bought a new pair of shoes, novelties for the children,* (likely gifts for her younger siblings. Evelyn's doll, that I now have, had been Edith's gift to her, likely given around this time. Resources were limited so all those little things made a difference to the family.)

*Regina beautifully decorated for WG Exhibition*[61]. (This was The World's Grain Conference and Exhibition, cosponsored by the City of Regina, the province of Saskatchewan and the Federal Government of Canada, and took place between July 24 and August 4, 1933, combining an academic conference and industrial exhibition with the annual summer fair.)

*Treated to supper with Julstads at Anderson's* (Likely the same Anderson family, who were original owners of the Victor Melby farm where I grew up. *Arrived home at 9:30 pm.*

***

That was certainly a busy month after Edith's year of teaching at Dravland. The next summer was different but still filled with an interesting variety of activities.

---

61    https://canadianstampnews.com/first-worlds-grain-exhibition-and-conference-opens-in-regina/

CHAPTER 14
# STILL MAKING GOOD WITH THE BAD

*1934–1936*
*Manitou Lake, Moose Jaw, Fort San, Minnesota*

*A Sunday Surprise party marked the beginning of the next summer, 1934 according to Edith's mini-diary. From L-R John, Walter Gronvold, Palmer, Evelyn, Dina, Mathilda with a wide brim hat, Lloyd in front and Clara behind, Edith, Adolph behind, Alvin Gronvold, Bernhard, Johnny with a lighter cap, Victor Melby, Clarence.*

*July 1-Surprise Mama and Papa to honour their Silver (25<sup>th</sup>) Wedding Anniversary, with a good crowd. Exciting ride in* (their cousin) *Alvin's old Ford. Johnsons stayed late.* (This was a delayed celebration, for John and Dina's actual wedding date was March 25, 1909)

This was another coincidence where a picture, matched up with Edith's diary, for that is their cousin, Alvin Gronvold and his truck, his mother, Mathilda, in the wide brimmed dark hat, beside Dina. The Hansen family is all there except for Anna and Myrtle, who were likely looking after meal preparation. I was surprised that my father, Victor would be there at this date for his folks had moved in 1930. I wondered if this was the time he was back at the vacated Melby homestead trying to make a go of it?

It was amusing that everyone's dressed in their 'Sunday-go-to-meeting' clothes while JB is decked out in work overalls, with his tattered straw hat in hand.

***

*July 2-Mama, Papa, Clara, Johnny and Mr. Opoien leave for Watrous. Rest up after many late nights.* Mr. Opoien was a bachelor neighbour with health problems, for according to page 121&122 of the community history book, the next year, he had gone back to the States to stay with family because of illness, and he passed away there. This must have been the first time the Hansens went to the great salt Manitou Lake, whose waters were reputed to have some therapeutic qualities.

*July 4- Begin teaching Gertie and Sam Koliger,* is written at the top of the next page, with ditto marks on all the dates, down to the 10<sup>th</sup>, where she wrote, "*Folks arrived home.* Trying to figure out what happened, was quite a puzzle, for I had spelled the name incorrectly. Finally, I found the story of Ben Koliger, their father, on page 249 of the local history book: Ben came out alone from Poland, in 1925. His farm was the piece of land south of the Dravland Cemetery.

Finally, and fortunately, in 1934, considering the historical events that were about to happen leading up to WW2, his wife, Pearl, and his two sons, Charlie and Sam, and young daughter, Gertie joined him. Now it made sense, Sam and Gertie were having English lessons, to be better prepared for school in September.

\*\*\*

*July 12- Left at 10 for Moose Jaw. Pleasant ride in Fred Preddy's truck. Reached Moose Jaw at 4 o'clock and attended the first session of the convention.*

They were at another Saskatchewan-Manitoba Young People's Luther League Convention at Central Lutheran Church in Moose Jaw. In the upper right of the long narrow group photo from this event, I spotted my uncles, Clarence and Bernhard-with their friends, Odin, James and Ken Torkelson and Sanford Johnson close by, Olaf and Virginia Ryan, Anna in the lower centre, sitting behind Alma Nelson, and Edith and Myrtle, towards the back right hand side. They were the only ones I recognized from this large group.

Besides convention sessions and taking part in 'interesting and helpful discussion Groups,' they visited Crescent Park, the Natatorium (Swimming Pool) and the Wild Animal Park and returned home from Moose Jaw on Monday, the 16th at 9:30 pm.

\*\*\*

Despite the tough times, the family took several family trips, like the one to Manitou Lake mentioned above. This next trip, I thought, happened after the young people's convention trip because Edith's little notebook now listed numbers, starting with Monday, the 16, all the way to 31, as if she had intended to keep records, but the page was blank, except for this note at the bottom of the page- *"31-Return home from Scobey,* [Montana] *10:30 pm."*

John had grouped all the trips he had taken together at the end of his memoir, but I moved this one to fit in with the thirties. John said- *In the summer of 1935,* [which I think was 1934, based on the observation made in the paragraph above,] *Edith, Myrtle, Clarence, Dina and I went to Fertile,*

*Minnesota. We went by Barton, North Dakota and visited Christ Holm and his family and had a pleasant time there for a couple days.*

*Then, we headed to Fertile, the year that a storm had destroyed many buildings in the area. We stayed with Mathilda, Jens Gronvold's wife and their children, who were staying with her father, Mr. Morvig during the summer holiday. We made our home at their farm and visited Dina's old friends and relatives. We had a wonderful time, for this was the area where Dina grew up. We visited Little Norway Church and the graveyard where Ragna Gronvold, Dina's sister was buried.*

*On our way home, we went to Hitterdal to visit my friend and schoolmate, Henry Brekke. One day, he drove us in a big truck, out to the Brekke farm, where I had my home in 1904, when I worked on a threshing rig. On the visit to the farm, the truck stopped in the middle of a slough, full of water. Henry got out in the water and carried Clarence first, and then me, to dry ground. [Did the road go through the slough, or what? In any case, Henry proved to be a Superman.]*

*After, we went to Detroit Lakes and visited Henry's father, mother and youngest son, George and family. We went through Fargo-Moorhead, about 30 years after I had delivered milk daily, selling it for 16 quarts for a dollar. We reached home all right and found everything in good order, after our first long trip with the old car.*

\*\*\*

*Because we had built up so well, had good water, and were near the highway and school, we waited and held out until the debt got so big, I thought I'd never be able to pay it. My brother-in-law, Adolph Gronvold, and Robert McNeil and I took a trip in 1937, to investigate places in the north. We travelled in our old two-door Pontiac car with a mattress in it, so all three of us could sleep on it at night. We visited several friends, who had moved up north earlier, and got many good meals but found no place we had any notion of moving to.*

*When I got home, someone said, 'What's wrong with you? You haven't enough money to move that big family!!'*

*A move like that- away from the old homestead where all of us both young and old had worked together for 29 years- would have been so hard.*

One of my mother's little history notes provided this explanation- "The wild oats seemed to frighten Father but the fact they had a good well on their home place was the deciding factor against moving."

\*\*\*

John continued, explaining a major health challenge the family encountered- *In the dry years, we had a newcomer who slept in the same room as Bernard and Clarence. He had TB and here Bernard got it and had to go to Fort San for two years.* [This was 1936. The TB Sanitarium was near Fort Qu'Appelle, east of Regina.] *They grafted bone from his leg into his lower spine. He has been well since. Later, Anna had to go to Fort San for six months, too, where her one lung was collapsed.*

It appears that there were several types of TB. In Bernhard's case, the disease affected his spine, whereas in Anna's situation, it was her lungs. My cousin, Larry Hansen said that his father, and my uncle Johnny, had been in the San for a short period of time in 1967. In his case, the TB had been latent for over thirty years.

John continued- *We made many trips up to Fort San to see them. Once, we were out in a terrible snowstorm, but we kept on till we got home.*

We can share some aspects of those visits made to the San, thanks again to some pages from Edith's notebook.

- On an Easter trip, April 16 &17, 1936, expenses for food, gas and accommodations came to $8.78 including two suppers at the San.
- An undated page indicated the Fort Qu'Appelle trip was 210 miles from home.
- They made four stops for gas and or oil on the trip up. (Weyburn- $0.15, Corinne-$1.75, Regina- $0.50, Wolsley- $1.30 plus $3.40 total for the return trip) came to $7.95, five meals-$7.05, rooms- Fort Qu'Appelle- $4.00 for a total cost of $19.00.)

These many entries lead me to believe that the gas tank must be small or the gas mileage poor.

Rooms probably were available for the San patients' family members, or places nearby provided accommodation. Since a few of the fuel entries I recognized as Clarence's handwriting, I thought this may have been the trip shown in the next picture, where they are having a picnic lunch.

***

*Picnic at Fort Qu'Appelle*

Clara and Edith are with their friends, Evelyn (front) and Alice Johnson (middle). They both worked at Fort San then. JB is sipping coffee from a thermos and Clarence is beside him on the opposite side of the picnic table hidden behind the paper bag. Besides the bag on the table, there are two gallon-size syrup pails that held the picnic lunch they are eating with obvious pleasure. (When I started school, a syrup pail like that was my lunch pail.) Edith took the picture.

# HOMESTEADING IN THE LAST BEST WEST

***

A letter from Bernhard provides an off the record observation of his Fort San Christmas celebration experience and shows a lot about the family relationships and how important letters were to everyone during that time.

*Fort San,*

*December 27, 1936*

*Dear Myrtle,*

*Was just wondering who I should write to but hit on you. Suppose I could as well have written 'folks', eh? Then it might be disappointing to whomever I addressed it.*

*Well, we are over our Christmas celebration, what with all the cards, letters, parcels and a good dinner I had. I should say, I had quite a good Christmas, considering. I want to thank you and Clara for the birthday book, Clarence for the cufflinks, (hope I can use them sometime) Johnny, Palmer and Lloyd for the monstrous Christmas card, all of you for the socks and bathrobe.*

*You, Mama, pulled a good one on me with that bathrobe. I wondered, how could it come from Army and Navy? There was no address on it, or postage and then the writing on it: "to B. R. H.* [his initials] *from us all?" I almost thought, you must have been up to Regina again.*

*Then, for cards and letters, I must have got from nearly everyone I know- 11 cards and letters at one time and four parcels, pretty good. I got a pair of slippers from Edith* (who was then out at the west coast), *a box of eats from Brown's, vue* (was that a picture?) *from Alma N.*

*I'll bet you don't know who I got a Christmas card from? Somebody at Weldon, I met in 34.* (This was likely the Luther League gathering at Moose Jaw previously mentioned. Several of the family, including Bernhard, had also been to the convention at Birch Hills in

1933, the previous year.) *I was almost going to say, to look it up in my hat but guess that's here. What mystery!*

*Well, I had in mind to tell you about our Christmas dinner. S'pose I'll do that before I forget about it. Here goes: To begin with our trays were brought in, covering them was a special paper and on it, was grapefruit, an orange, a little green paper bag of candy, with the handle decorated with stars, a nice Christmas card, a box of chocolates from Lokken's with my name on, a sprig of spruce, and nicely coloured Christmas paper serviette and of course salt, pepper, milk. That was the first course. The grapefruit dishes were gathered up and then we got soup. When that was through, they brought in the turkey, etc. After plugging that away, they rounded off with a plum pudding and coffee, the first really good coffee I've had here.*

*I've really got some writing to do, to answer all those I got stuff from, or heard from. I'm going to try to answer as many as possible before 1937 so that I can use the Christmas seals I have left.*

> Christmas Seals (Stamps to decorate Christmas card envelopes) were used to raise money, initially to provide sanitariums for those suffering from TB and later provided funds for X-rays and tuberculin skin testing. The design of the 1936 stamp was a boy and girl in profile standing on either side of the double barred cross, the symbol of the Lung Association. My mother always used Christmas Seals on the envelopes of the cards she sent out each year when I was at home.

*Oh, yes, thanks for the Golden Memories request. I heard one a few days ago, but didn't know if it was the same one or not.* (This was a radio show, where listeners wrote in to ask to play a song dedicated to someone for a special occasion.)

*Well, was Santa good to you?* (Not that the family ever put much stock in Santa, it was just a figure of speech. Gift opening in the Hansen Family was on Christmas Eve, according to the Norwegian custom.) *S'pose you're having considerable holidays.*

*I happen to remember Johnny asking me about stamps* (He collected stamps.) *I'll try to remember to put in a sample when I finish this letter*

*I'm just listening to this Chase and Sanborn's program with the actors* (a radio comedy and variety show sponsored by Chase & Sanborn coffee) *and listening to Mr. Christian here, rant about how silly it is. That's the program that takes the place of the Goodwill Court.* (This was a popular human interest radio program from the mid1930s, its final episode was December 1936.)

*This is in to fill space, I got off the subject as you see. In the afternoon, we got nuts, grapes. Couldn't make much of supper after that.*

*Have you got any snow down there? We had some today here, just so you know it snowed.*

*Well, I can't make a connected story trying to listen to the radio, so I might as well just end. Hope you're all fine. I'm quite well and hope to hear from you soon. Thanks again for everything for Christmas and I hope you have a very good year in 1937. Sending you this piece of Spruce for remembrance.*

*Bernard*

***

The Thirties were the big part of this Hansen Family story when most were in their teen and young adult lives. John didn't provide any of those details, so the mashup of notes, letters and photos collected here, show us what life was like for them.

And now before we go on a west coast adventure, with Edith on her year off visiting her uncles, aunts and cousins who are now living in Seattle, Washington where, we'll see some interesting comparisons of the costs of common daily items and ride along on a 48-hour trip from the west coast straight through back to their Saskatchewan homes, we'll fit in a story my cousin, Larry shared. His father, Johnny talked about how he and his younger brothers, Lloyd and Palmer used to play with their toy trucks and tractors way down in the farthest corner of the trees. If they were far away, they wouldn't hear when JB was calling them, for then they would have had to go and work.

CHAPTER 15

# STILL AWAY FOR A WHILE

*Seattle and Vancouver*
*1936–1937*

One year of my mother Edith's working career decade, was spent away from teaching. A one-time conversation, Edith had with my sister-in-law, Sheila, indicated that perhaps John had advised his daughter, Edith to take time off, for the sake of her mental health. Her photo album showed a prairie girl on a skiing expedition with her west coast cousins near Seattle, Washington and being part of several functions, such as the wedding of her cousin, Albert to Ella-Mae, one of the first of that generation of the Gronvold family to be married.

Edith's well-worn Moyer's School Supplies Ltd[62] notebook, smaller than my cellphone screen, clarified parts of that trip. Edith travelled out to Seattle with her Uncle Henry's family. As she calculates the daily expenses (or her heading, 'board and room for the trip'); each of the costs are divided by five. That was her uncle Henry and his wife, Helen, their two daughters,

---

62   On the cover: "Canada's School furnishers. The pocket booklet is sent to you with our compliments that you might become more familiar with our New Name which we feel is more descriptive of our business." A 1935 Calendar is on the inside front cover. This may have been a freebee given out at Teacher's Convention.

Irene and Martha, and Edith. The total came to $3.50, and she recorded, "Paid Henry $5.00."

As Edith starts out on the trip she lists *"$25 on hand: Head Tax- $8.77, at Westby, Montana and overshoes-$1.19."* [It's the end of October and I'm guessing winter weather has arrived. The tax was because she was staying in the USA for more time than would be considered a holiday.]

The notation, "Surprise Party gift of $10" and a "$25 check from her uncle Olaf on Dec 10 and Feb 1," made me wonder where and when this had been? But in any case, her uncle was making sure she was taken care of. She hadn't had a lot to start out with. Her total recorded earnings from Shakespeare School, the year before she took this trip, came to only $129.50 and out of that she still had to pay board and room.

***

While Edith stayed with Jens and Mathilda, she sang in the choir at their church. The Christmas Day service was broadcast on the radio, and the Hansen family, back in southern Saskatchewan, picked up that station on the radio and heard the choir perform.

## Shopping Across the Line

48 hours was the minimum amount of time you had to be across the border to bring back items without paying duty. I remembered, "taking a 48", became a yearly habit for my mother and my aunts to do the needed shopping for the season, stocking up on cottons, jeans and shirts and sheets, particularly in the fifties when I was growing up on the farm. And there were times they stayed with their cousin, Martha, who lived in Plentywood.

I loved those tangerine jeans Mum bought, the same vintage as the rust and turquoise fabric she selected to make the peasant-style flared skirt and top I wore in the 1956 family portrait.

\*\*\*

Twenty years earlier, her notebook revealed the shopping she did:

*Received from home–$7.00*

*Bought dress– Mama –$ 3.70*

*Stockings– Mama– $0.49*

*Stockings –Anna– $0.49*

*Dress –Anna $1.00* (Anna would have been about 19 and Clara, 17.)

*Dress –Clara $1.00*

*Stockings–Evelyn $0.29*

These were examples of how the family took advantage of more shopping choices in the States.

Edith's personal shopping beginning November 4, 1936, was interesting to note. Sixteen items listed under 'Expenses in Seattle', were mostly Christmas gifts of 10 to 30 cents each. The most expensive was $.69 for a tablecloth for Mama and that same amount, for the slippers, Bernhard acknowledged receiving, in his letter from the San, recorded in the previous chapter.

Three times in 1937, between June, and when she made the return trip to Saskatchewan in August, she had rubber lifts replaced on her old brown, new brown and gray shoes, all to keep what she had in good repair. Corn plasters, listed three times, indicated one or more of her shoes is causing her trouble. A new pair of shoes for $7.50 seemed like a big price for that time, but she had long narrow feet and always paid more to get properly fitting shoes.

When Edith's time in the States was up, her Uncle Jens and Aunt Mathilda, accompanied by their daughter-in-law, Inez and her mother, Mrs. Sather, drove Edith back across the line to Vancouver, and did some sightseeing at Stanley Park.

The story gleaned from Edith's recording of her Vancouver expenses began April 24, 1937, when she placed an ad in the **"Province"** newspaper and rented a room from May until August 5. All of this cost $18. She got a position doing housework, for $5 a week, starting on Monday, May 3. She may have worked at more than one home because her book had several names and addresses of people, I thought, were prospective homes and/or employers.

Bessie Volen, and her sister Marian, were Saskatchewan friends, taking a reprieve from the prairies. She chummed with them, as well as several others she got to know while she was in Vancouver. Bessie had been a fellow teacher friend, maybe from when she attended Normal School[63] in 1929. Bessie had taught earlier at Greendale School, (Edith taught there in 1939-40.) Several west coast outings show up in her photo album and expense record: a trip to Victoria, picnics at Kitsilano and Second Beach, near Stanley Park, a July 18 trip to Capilano, and a picnic with the Orr's, another transplanted Saskatchewan family who had lived at the farm a mile east of our Melby farm, in the early years.

***

Six made a return trip back to the prairies, via the US route, including Edith, Bessie Volen, her sister, Marian, and three fellows, identified only by their first names- Marvin, Tiny and Fred, though I suspected one of them might have been their brother. Ten stops for gas from Vancouver to home in southern Saskatchewan came to $16.40. They came via *MacDonald Pass 6225 feet* [the Highway to the Sun] *10:25 am Friday,*

*Stopped at Helena. Dinner at Great Falls.*

---

63    We would call that Teacher's College.

*Havre at 6 pm, some rough roads, 2 flats between Malta 9 pm and Glasgow. Got a new* (inner) *tube in Glasgow. Repaired other old one. Left at 1:15 am.*

*Took our time, as we wouldn't be able to get through Customs before morning. Tiny gets his way and goes via Williston, Fortuna, and then over the roughest roads to Ratcliffe.* (That town was about 5 miles southeast of the Hansen farm, and was the next town west of Hoffer, on the rail line.

[The customs officer, Mr.] *Topping was very nice, didn't even ask to see our suitcases. Likely the only bright part of our God forsaken Saskatchewan, as far as Tiny and Fred could see anyway.* [We] *arrived home, 9:15 Sat morning – 48 hours after leaving Vancouver. Good going!* [They had] *breakfast at the Hansen place, took pictures and said their goodbyes."*

Based on the fact, Edith paid up rent until Thursday, August 5, the group left that morning, drove straight through, despite stops to fix several flat tires, she made it home for breakfast Saturday morning, August 7, 1937. That's a surprising 48-hour trip!

\*\*\*

Later "in 1937, the last 10 miles of Highway 35 connected with the US Highway 85 [at Fortuna], south of Oungre, and a new Customs-Excise office and residence building was completed and the Ratcliffe office closed."[64]

\*\*\*

Did Edith start something? A decade later Lloyd, Johnny, Anna and Evelyn, and Mrs. Evelyn as well as Bernhard, spent a period living in Vancouver and of course, Clarence had his time of army training on the west coast.

\*\*\*

---

64   **The Saga of Souris Valley RM No.7** from The Topping story written by Tommy Topping Page 400

Before leaving 1937, an item worthy of special note was a Christmas gift and letter that the Hansen Family received from the Max Frohlich family, their Jewish neighbour who lived a mile west. The memory of this letter had remained so riveted in my mother, Edith's memory, I knew I would mention it. During the writing of this manuscript, I had an opportunity to share my mother's special memory via email, with his son, Chuck Frohlich. He said his family had kept a copy of that letter which he thankfully agreed to share. What a special touch that added to this story!

> Dec. 24. 1937
>
> Dear Mr & Mrs Hansen: —
>
> Please accept our little gift for Christmas as appreciation for your friendship.
>
> We are sorry, that two of your dear children will be missing at your Christmas dinner this year.
>
> It is our hope and wish that at the next Christmas the Lord shall bless you with the pleasure of having all your family in the best of health at your home for the holiday season.
>
> Wishing you all the very best for the season 1938.
>
> Sincerely yours
>
> M. Frohlich & Family

*Frohlich letter*

When I realized, this letter is dated December, 1937 and it's been a whole year after Bernhard's letter recorded in the previous chapter, and now refers to two from the family being away, I began to understand how long and drawn out this period of time had been for all concerned. (Anna came to the San the fall of 1937, and stayed about six months.)

***

Now John's memoir picks up the story. *After Bernard got home, I asked him if farming wouldn't be too hard for him and suggested he go to normal school to be a teacher. He did and taught a few years and then decided to go and take a naturopath course in Chicago.*

*Since Bernard and Anna were both healed and haven't had to go back, we have a lot to be thankful for. Look at the hospitals and finally, to God, who again and again, has been so wonderfully good to us and our children.*

> Bernhard practiced out in New Westminster, BC for a time. Before I started school, he sent me a postcard I saved that showed the hotel and business section of Vancouver, BC. He marked with an 'x' where his office was. And on the back he wrote, "Did you have a nice Easter? How is Cookie? Do you have the cat yet?"
>
> So, was Cookie that much loved black cocker spaniel Bernhard had given us back then? What a sad day it had been a couple of years later, when we came home from school to find out Grandpa had been out at our farm that day, and in helping with cultivating around the garden, our dog, we called 'Patsy,' had been run over.

Now that we've been out and about, an eight-year old's impression of the depression takes us back to more of the everyday routines on the Hansen farm.

CHAPTER 16

# BACK TO LIFE ON THE FARM

*Views from the youngest*
*1937*

And now back to the realities of farm life, by way of youngest brother, Palmer's contemplative observations. My mother, Edith received this June, 1937 letter about five to six weeks before she returned home from her escape to the west coast.

*Hoffer, Sask.*

*Thursday, June 1937*

[Though no date was given, June 17 seemed most likely, based on what he writes.]

*Dear Edith,*

*I haven't had a letter from you for a long while now, so I am writing you a little, so it will make you write a letter to me. (You can't expect a letter if you don't write one!)*

*Johnsons went north to see the land last Wednesday and came back today. I think that we might see Johnsons out there some day this year.*

(The Johnson neighbours moved to Carievale, in the southeast corner of Saskatchewan.) *It sure is nicer* (in that area) *than the country here now. It is just wind blowing now, with all the dust, too.*

*We might come out there sometime in July, if we would go out there. We won't go yet for a while. Clarence is down to Outram, where a man has a shop. He went to fix up the car, so we could go to British Columbia in it. He might come back tonight sometime.* (Edith's notebook indicated she had researched alternate ways to take the trip back to Saskatchewan. The (train?) fare quoted to Culbertson, MT was $21.67 and to Weyburn was $30.85. And it appeared from Palmer's letter that taking a family trip out west must have been discussed.)

*There are gooseberries and there are green strawberries and flowers on the raspberries. There were some cherries had flowers, that had fallen right off. The trees keep green in this dry country, it is funny that they can keep green, when it is so dry. There has been no rain here at all. Have you had very much rain out there lately? There has only been wind, dust and weeds all around here. The radio only brings some static to hear now today.* (I am guessing, based on the fact the farm had a good well, an important chore was carrying pails of water to irrigate these berries, and get their vegetables to germinate and grow.)

*Do you know, we have a bicycle? Lloyd is chasing the cows with the bicycle. Lloyd went over to Frohlich's on the bicycle on Tuesday evening, after supper, and he went to Elfenbein's and Hvidstones last evening, to see if the bull was there. He went with a bicycle, too.*

*It is a second-hand bicycle, bought for $12 cash. It had a good back tire but not such a good tire on the front wheel. We got it on June 11, in Estevan. The next morning, Anna was going down the hill, she bumped into the fence and broke the front wheel off and she was scratched on her arm and on her leg.*

*How much do you weigh now? I bet you weigh 150 pounds now. Do you? It is about 5 o'clock now. How do you like British Columbia, Vancouver and all the country out there?*

*Each person got a coronation badge at the picnic* (the community school picnic we learned about in Chapter 9. This was the year, Edward abdicated, and the event of that year was the coronation of King George and Queen Elizabeth.) *Well, you missed the picnic.*

*Grade 4 to 7 are having exams on the 23rd or 24th and 1, 2 and 3* [have classes only] *until dinner, the 29 and 30.*

*I sure wish I would pass to grade 4, so I could have geography. I would like geography.* (Edith had been his teacher two years before in 1935, when he was in Grade One. In a one-room school, you are very aware of what the older grades are studying.)

*I guess I'll write some more again in some other time, because I haven't anything to write in a letter.* (This letter was written in installments.)

*Papa was just south with Webster* [his stallion] *to breed horses and he just came home right now.*

*I have had an earache since Saturday, but it hasn't been bad since Monday now.*

*Have you lots of fun out there now? Lately, I couldn't get much fun because it was too dusty, and it blew too hard.*

*We get four pails of milk each day and about 1 gallon and quarter sometimes. We milk eight cows now, and we've got 17 cows all together.*

*Are the people very fussy, out there now?* This must be referring to something that Edith had written about her job in a previous letter to the family

*Grade 4 are making tea towels and Grade 5 and Grade 6 are making cushions and Grades 7 and 8 are making dresser scarves for to get money to buy curtains for the school.* The materials they used for this project would have to have been the muslin fabric saved from the 100 pound flour or sugar bags. Once those bags were empty, the chainstitching would be unraveled, opening a flat square of cotton fabric that would be washed and bleached. (The string would be wound up into a ball that was saved for other purposes.) That material, when it came in printed designs instead of the usual unbleached muslin, was also used for making items of clothing. But no other kind of tea towel compared to one that had originally been a sugar or flour sack.

*** 

*I must hurry up a little because we have to send this letter away today, and I have to go to school, too, so I guess I better quit writing. Please write.*

*Your friend,*

*Palmer Donald Hansen*

## Milking cows

Every morning and evening, milking the cows was a big part of daily chores on a mixed farm. A notation, 'Cream Check- $5.43," in Edith's notebook, was evidence they likely shipped a three- or five-gallon pail of cream to the dairy on a regular basis, when the biweekly train made its return trip back to Estevan. Though the amount earned was small, it was still cash coming into the household. Meanwhile, I'm sure there was still plenty of cream left for the other things needed for feeding this large family.

Another page of Edith's notebook listed:

1. *Float for an Economy King Separator*
2. *Spring in handle*
3. *Rubber Rings.*

These were all parts of a cream separator, and this could have been a reminder to pick up on a trip to a bigger centre, or more likely to order from the mail order catalogue.

Once the fresh milk is brought into the house, it is strained into the top bowl of the separator, then "the person operating the separator...turns the handle around and around and the machine builds up centrifugal force as it spins, thousands of RPMs develop, causing the milk to be pulled against the walls of the separator while the cream, which is lighter, collects in the center."[65] Skim milk was collected from the bottom spout and cream from the top spout.

Cleaning the separator used to be my least favorite dishwashing task, when I grew up on the Melby farm, especially dealing with the stack of cone-shaped discs from the centrifuge to keep it fresh and clean. It had to be done as soon as possible after finishing otherwise the job was more difficult if the milk soured.

Separating, as described above may have been done only in the mornings or after the evening milking and in the off times, milk was just strained and cooled in containers to be used as whole milk. In that case, the cream rose to the top and had to be stirred in, to mix it or it was skimmed manually with a big flat spoon, if you didn't have a separator. Just pure raw milk, it was not pasteurized or homogenized.

With a large amount of milk and limited refrigeration, some of the milk likely did get sour. When that happened, the sour milk might be used to make pancakes or bread or they more likely set the pot of it at the back of the stove where after slow and gentle warming, the milk would separate into curds and whey. The solid part strained out was cottage cheese and the liquid whey was fed to the pigs or chickens.

---

65    http://antiques.lovetoknow.com/kitchen-collectibles/antique-cream-separator

*Clara and Johnny doing chores. (From a group of pictures in Edith's album, titled "Johnny's pets" In the background is the Avery tractor, the well drilling machine is laying down to the left of it.)*

## The Plentywood Cookbook

Cream that got sour might be used to make the cream cake recipe from the **Plentywood Cookbook,** a much used cookbook in our house, as it likely also was on the Hansen farm in the forties. I remembered the Sour Cream Cake, having made it more than any other:

### Sour Cream Cake

1 c. sugar,
2 beaten eggs,
1 c. sour cream (today you could use a full 250 ml container)
1 2/3 c. flour, sifted with 1/2 tsp. soda,

1 tsp. baking powder,
1 tsp vanilla.

Mix in order given and bake in two 8 inch layer cake pans or an 8 inch square pan at 350 degrees F until done.

Since a square pan doesn't go that far feeding a big family, the recipe was usually doubled and made in the larger 9x13 inch rectangular pan.

\*\*\*

Barb Raaen was an online friend from Minnesota, who helped with research on my first book, for she had family from Plentywood and shared the Gronvold history from that Divide County community history for my first book. Then, when I wanted to refer to my worn-out **Plentywood Cookbook** for this story and I had no idea where to find it after our 2017 move, I emailed Barb to ask if she had a copy. [66]

She replied- *Years ago, I had a beat up Plentywood Cookbook, which I gave to someone in my Montana family. However, in 1999, they reprinted it as a double – on one side is the old cookbook, flip it over and there is a second cookbook called Turn of the Century Cookbook, Celebrating the New Millennium. The original had been put together by a Plentywood WMF group* ['Women's Missionary Federation', another version of 'Ladies Aid'] *at the end of the thirties. The cookbook was printed in 1937, 1939, 1940, 1941, 1942, 1943, 1944, 1945, 1947, 1948, 1949 and 1999.*

*I found four recipes submitted by Mrs. J. C. Gronvold, Seattle: page 24 – white cake, page 44 – chocolate coffee frosting, page 118 – liver sausage, and page 171 – potato salad & dressing.*

I picked the recipe, **Potato Salad & Dressing** submitted by my grandmother's sister-in-law, Mrs. J. C. Gronvold, from Seattle because it uses four staple items produced on the farm: potatoes

---

[66] Emails from Barb Raaen- Nov 30, 2017, Dec 1, 2017, Jan 13, 2018

and onions from the garden, with cream and eggs used to create the salad dressing

*Potato Salad and Dressing*

*4 eggs*
*1/2 c. vinegar*
*1 tsp. salt*
*1/2 tsp. [dry] mustard*
*1 pt. sweet cream*[67]
*1 c. onions*

*Dice cold potatoes. Over the potatoes pour the dressing made as follows: Beat the eggs to a froth, and add salt and mustard dissolved in vinegar; add a pinch of white pepper, if wanted. Set the container in a pan of hot water, stirring frequently until thick.* [This could be done in a double boiler.] *Pour in cream (whipped, if desired),* [but in that case the dressing should be cooled first] *to thin it out, and add the finely chopped onions.*

(Sadly, shortly after Barb's help with this, I learned via Facebook, that she had passed away.) I am thankful for her help and the encouragement she provided on this book as well as my previous project.

Cream was the precursor of butter. Some churns were small jars (possibly 2 quart size) that had an eggbeater-like mechanism to agitate the cream until the butterfat glommed together, but most likely, I remembered a large metal churn that held a gallon or more of cream with big wooden paddles that sloshed the cream around until the butterfat solidified and the remaining liquid was drained off. That by-product, buttermilk was used as a beverage, or again for making pancakes.

---

67    A US pint is 16 ounces, or 2 cups or 500 ml is close to that amount. In Canada, an Imperial pint is 20 ounces, or 2 ½ cups, so the Imperial quart and gallon were proportionately larger.

At our house, a staple Saturday night supper meal was pancakes with corn syrup. I wonder now if that menu item was a carryover from the Hansen household.

Another common basic Norwegian item was *Risengrøt* or Rice Porridge. It took several hours of cooking in the oven or in a double boiler and used lots of milk (4 cups to only 1/3 cup of rice) and didn't require much attention other than giving it a stir every now and then. My mother said when they had *grøt* that would be the entire meal. It was served with a bit of brown sugar, cinnamon and melted butter. But I was surprised that *rømmegrøt* or cream porridge made with cream and flour, which also was served in that same way, was not one of the dishes they made, considering they had a good supply of cream. I learned about that treat at the Sons of Norway Norwegian Cultural Camp.

Cream, like the whipping cream of today and sometimes even thicker, was what was used in coffee, and a cream pitcher was always on standby to pour that richness over canned or fresh fruit or puddings.

When the first of the tomatoes ripened on the vine, I had seen my grandfather take and chop up one of those red beauties into a cereal bowl, sprinkle it with sugar and drench it in cream. I think that was a favorite of his. Try it sometime.

And another farmhouse treat was fresh leaves of garden leaf lettuce, tossed with a homemade dressing made with cream, augmented with a splash of vinegar and a sprinkling of sugar, salt and pepper. Simple, and oh so good!

\*\*\*

So, though times were hard they always had enough to eat. Rains eventually came; the Dust Bowl of the Thirties was relegated to memory.

CHAPTER 17

# READY TO LEAVE THE THIRTIES BEHIND

*A time of transitions*

Now that Edith was back in Saskatchewan, where was her school for the 1937/38 year? It had to have been Workman School near Carievale, SK, based on the fact she taught there after the Leonard Johnson Family moved. However, other notes of hers, in two different places, indicated she taught at Blooming for two years, 1937-1939. Pictures in her albums and addresses on a stack of Edith's old love letters gave proof to the Blooming location, for romance with Victor Melby was developing in that period.

At a time, earlier in the thirties, when the local young people's group had taken a trip to Carlyle Lake, Victor Melby had been the truck driver. Edith admitted to me, one time that she was jealous of the fact that Olaf Ryan and Virginia Bugg, who were courting at the time, had the opportunity to ride together in the cab while she had to ride in the back of the truck with the rest of the crowd. Since Olaf and Virginia[68] were married in December of 1935, this trip could have been earlier that year or possibly even 1934. In any case, economic conditions managed to extend the length of Edith and Vic's romance.

---

68    Lloyd Hansen married Lauretta Ryan, daughter of Olaf and Virginia.

\*\*\*

The year before Edith was married, 1939-1940, she was at Greendale School in the northwest corner of the Souris Valley municipality. It was there she received this puzzle letter from her brother, Bernhard. The end was the beginning, and the sentences alternated between being written backwards and forwards, and it read from the bottom to top of the page. I thought he may have been trying to copy what his Jewish friends learned- for in Hebrew the beginning is the back of the book.) Though there was no salutation, it was sent to Edith.

At the time Bernhard wrote this, he and his Uncle Adolph were farm-sitting at his Uncle Henry's place, near Plentywood, MT. The Jens Gronvold family were already established in Seattle, Washington as was Olaf Gronvold, and Henry and Helen were considering it. Bernhard and Adolph, had just returned from a trip out to Seattle.

March 23, 1939

*I'd promised you I'd not write, till you acknowledged receiving the parcel, but I guess, I will not wait till you get back for Easter* [April 7-16, 1939] *to tell you the little Seattle news, as you call it, that I got. We got back fine that day, but there sure was more snow here, than out there. It had almost all thawed when we left, so I suppose now, they will almost be having summer.*

*First, I can assure you, that you won't get a ride to Fort San at Easter, maybe you won't want to now, as Anna is home anyway.* [She was in the San for six months.] *How about Regina or Plentywood? I believe, that's too early to go to Plentywood, as we had planned on going there sometime when the weather and roads are good, to do some work on the car.* [Have you noticed how much travel was affected by the weather and the condition of the roads? And even though the places where Edith was teaching were only 10 miles away from home, it appears she only got home for the major holidays, not weekends.]

*Boy, has the weather ever been nice, the last three or four days. There is not enough snow for sleighing, and wheel traffic is almost out of the question, too. I was driving some colts yesterday, and in places they fell into the snow where it was 2 to 3 feet deep, and then again, some places, it was so bare that the sleigh really pulled hard. About our stay at Henry's, that was just swell for us. I don't know whether the place suffered too much. We were to Plentywood three times with Espeland, and only once, was it hard to get through with the car.*

*We, Adolph rather, had only two cows to milk and only one more critter all together, and less than 50 hens. That's not many chores in any man's language. I did the cooking and I thought, I really had something in potato boiling. I used to boil them with skins on, a kettle full at the time. And boil them over again, what we used every meal. They're kidding me here about that, because I thought it was a good idea! Enough about that.*

*Helen, as you may know, is staying in Plentywood at a hotel with Martha. As to Seattle news, she was kind of peeved with Harold* [their son-in-law] *for wanting to quit his WPA job and wanting to just raise chickens for a living.* [The WPA or 'Works Progress Association' was part of FDR's Public Works program, instituted in 1935 to get unemployed people working.[69]]

*They were all very well, but Henry anyway, didn't like the rain. He did talk a lot about moving out there, for the winters anyway, though. he thought it was drier there, than here, in the summer months.* [Yet that doesn't seem right.] *Helen, I guess would like to move out. Irene* [their daughter] *lives in a so out-of-the-way place, they couldn't drive all the way up there, on account of the poor roads, or the wet.*

*Henry seems to think* [his son-in-law] *Harold wasn't so bad, he admired his electrical ability, and said that he would like to learn electricity himself. Taylors were still at Jens's and Jens had a lot more business than* [his brother] *Olaf now. Mabel was getting married to*

---

69   http://www.livinghistoryfarm.org/farminginthe30s/money_06.html

*Ralph in the spring sometime (may not be much news to you). Her folks were coming out for the event.* [Probably more prairie people they knew, for besides moving up to northern Saskatchewan to escape the Dust Bowl conditions there were also several former Souris Valley residents who moved to the west coast, some to Washington state or the lower mainland of British Columbia.]

*Well, I know of no news to tell you, so will just ramble off to get my fish in out of the warm weather. I don't know whether it was you, or Myrtle that had asked if they could come for you by car, but I expect so, if the roads are OK and it seems they might be, if this weather continues.*

*Bernhard*

***

I knew there had been an electric generator in the basement at the Hansen home for I remembered they did have electric lights before rural electrification came to the district, in 1952. Pictures show the tower, connected to that system, on the northeast corner of the house. All this they got from the Henry Gronvolds, when they moved off their farm, probably around this time as the Forties begin.[70]

***

Edith and Victor's courtship is gaining some headway. In a December 1939 letter, Victor expressed his wish that Edith would be able to get a school up in the northern area around Birch Hills or Weldon.

Job opportunities for Vic were few and far between, although he cut firewood over several winter seasons. Because his father's health was failing, he often helped him out in his door-to-door grocery sales business.

---

70   Email from Larry Hansen Sept 21, 2017 from information his father, Johnny shared.

Henry Gronvold's address on Edith's October 1940 wedding register was still given as Plentywood.

The effects of the war underlay everything, as Victor writes, *"Oh, if I could think of a good excuse for you coming up here, wouldn't I be happy but Edith Dear, we will have to* [carry on] *for there are many people throughout the world away from their loved ones, without any hopes of things brightening for a long time to come."*

Around Easter time, he wrote, *"I hope I'll have a job in sight by the next letter, as things should start to move here pretty soon now but I guess it is no use looking for anything very big this spring with forty cent wheat in the fall."*

\*\*\*

As the Forties begin, farming conditions improved and now the world is at war.

A February to October portion of a 1940 calendar, included in Edith's wedding scrapbook must have marked her official engagement to Victor Melby, I thought, for soon the JB Hansen farm hosts a wedding celebration.

# Melby/Hansen Nuptials[71]

A wedding of interest to a large number of Saskatchewan friends took place on October 26th, when Victor Rudolph Melby of Birch Hills and Edith Marie Hansen were united in the bonds of holy matrimony at the home of the bride's father, Mr. John B. Hansen, of Hoffer, Sask. Rev T.J. Langley officiated.

The bride was attended by her sisters, Myrtle and Anna Hansen and the groom by Palmer Melby of Birch Hills and James Torkelson of Beaubier. Before the ceremony, the audience united in singing, "O Perfect Love." Then the bridal couple entered the beautifully decorated room to the strains of the Wedding March, the bride wearing white satin and a veil. The wedding service was followed by several speeches and musical numbers, much appreciated by the large number of people present. A delightful buffet luncheon was served.

The happy couple, who will make their home at Tisdale, Sask., received many beautiful tokens of high esteem. The bride has taught public school in the southern part of the province. The groom is a travelling salesman. May God's protection and blessing rest upon the new home.

---

71   The source was not given but it was most likely the weekly paper, **Estevan Mercury**

CHAPTER 18
# TIME FOR CHANGE

Edith's wedding scrapbook opened with the preceding newspaper write-up of their Wedding. Since everyone is still recovering after the decade of the depression, this provided an example of how to have a wedding without spending a lot of money.

Edith's ivory satin dress was ordered from the Eaton's catalogue, the bridesmaid's dresses were made by Anna of sky blue taffeta for hers and pink taffeta for Myrtle's. It was a full house that day, with 105 people listed on the register. 56 invitees were listed who couldn't come.

The scrapbook showed over half of the cards were simple miniature gift cards, the remainder were meaningful personal sentiments, written out by hand on plain pieces of paper. Their studio photos taken a month and a half later at Tisdale, SK, came to $11.73. (That was for a dozen in folders for gifts and a dozen plain.)

Seven waitresses decked out in white and pink crepe paper aprons were her sisters- Clara and Evelyn and friends- Evelyn Johnson, Edith Torkelson, Ruby Johnson, Herborg Roen and Vivian Fossum. The menu they served included brown bread cheese sandwiches, buns with chicken and ham, jellied fruit salad, chocolate pinwheel cookies, Swedish drops, two-toned walnut cake, dill pickles and olives, coffee and orange nectar. The wedding cake made and decorated by Anna, weighed 13 pounds and was reported to be delicious.

The wedding gifts of note were Myrtle's gift of $150, and $55 from nine money gifts of $1, $2 and $5 amounts and a table made by Mr. Juel Marken. (All my years growing up, Edith used it to display her geraniums and Christmas cactus, while the shelf below held farm newspapers and magazines.) Marken had said, in Norwegian, which somehow made it sound comical- 'when he'd make one for Myrtle, it would have a drawer.' This was a story; I heard my mother repeat many times. But Myrtle remained single all her life.

***

Once the festivities are over, Edith shared in her scrapbook, "October 29 at 2:30 pm, we are on our way to "our" home. The little old Ford car can carry quite a load. Look! (She might have been referring to a picture of herself getting into the car.) Stopped at Corrine for a short while, then on to Regina where we took in the show: "North West Mounted Police" at the Capital. {This might have even been the Premiere of this movie which opened in Regina on October 21, 1940.] Spent the night at Hotel Champs. Shopped around at stores for furniture but bought none."

Finally, October 30, they spent the night in their first home in Tisdale. And the next chapter of family life begins.

It would be nine more years before the next Hansen family wedding took place:

    1949- Clara and Harald Tangjerd,
    1950-Johnny and Evelyn Erickson,
    1951-Clarence and Vera Tedford,
    1952- Anna and Olaf Hagen,
    1956-Palmer and Marg Rogers,
    1960-Lloyd and Lauretta Ryan.
    And in 1981, Bernhard married Hazel Sheske.

***

After this special event, we get back to more of the daily life of the forties, thanks to letters Clara received from her teenaged brothers' from back at home while she was away for job training and work.

## Chapter 19
# FORTIES ON THE FARM AND BEYOND

*Early 1940s*
*Letters tell the story*

The life and times of the early forties was best described by several significant letters my aunt Clara had saved. This first one was a joint effort between her younger teenage brother's, Johnny and Lloyd. With sassy extra messages inserted between paragraphs as well as along and around the margins, the note queried "Have you become curious or dizzy yet?"

*Dec 1, 1941*
*Dear Clara,*

*Here I am again (maybe at last). I don't know how much news I have to write, but I'll try.*

*It is very nice weather out now after a wintry spell. Today it was melting all day and still is tonight, about 35° right now.* [Freezing was 32°F] *The snow has sunk a lot today.*

*The roads were good, opened up yesterday by a bunch of guys from Ratcliffe* [a town about four miles west] *with two cars and a tractor. They had to detour half of the way almost, today we have the Essex*

[car] *out and we went a mile or two each way, just for fun. Several cars went by today.*

*Sam Frohlich* [one of the older sons of their neighbor, the Frohlich family] *is going to Toronto the day I mail this. I guess he will have a nice trip by train.*

*We butchered 16 roosters today, some are mine,* [and] *we also killed Lady* [a hen, they had named?]. *They are going to Torquay to sell them tomorrow.* [That would have been a busy day at the farm with plucking feathers, singeing, eviscerating and cleaning the birds so they'd be properly dressed for sale.]

*Papa may also go to Estevan because he says he has some teeth trouble.* [At some point, JB got false teeth, this might have been the start of it. In my memory, he always had false teeth.]

*Johnny & Clarence are trying to send in an advertisement to the Western Producer* [a weekly farm newspaper] *to sell our feed grinder and the Essex* [car].

*It sure is nice weather out now. (Nice and moonlight, etc.) Evelyn was home last weekend, the first time for about three weeks.* [This was likely the year Evelyn took her Grade Eleven at Lake Alma.]

*We were skating a few days ago and had a hockey game. The ice was very rough, but we flooded twice so it is nice now. It has melted since and frozen so* [it's] *quite smooth now.*

*Dravland* [School] *still has Mrs. Schreder as their teacher. She will leave at Christmas, if not before.* [She probably was pregnant.] *The school board got four teachers who answered the ad, one was an old man, 64 years old, one was a married woman and the other two were not married, I suppose but the* [fountain] *pen went dry so I can't finish.*

*CKCK Regina-*[radio station] *news of the war, Libya, Russia, Japan, Canada farmer's war, USA miner's war.*

*We didn't get this sent before because we had no envelopes.* [How many times have you seen that excuse?]

[Note written crosswise over the first page] *We were to Weyburn last Friday with the chickens and turkeys, the road was rough. I just about lost seven dollars, too.*

[Another side note indicates they had been into trapping] *Fur market: rabbits- $.50, skunks $2.50. Muskrats $4.50*

*December 8, time to go to bed. Mama and Anna are ordering Christmas presents* [from the Eaton's or Simpson's mail order catalogue.] *Maybe I should send for some, too. Do you want any?*

*Your pig is getting along fine. We have the pigs over in Buddy's barn now (about 70 of them) including yours. Your biggest one was sick the other day, almost croaked, ha ha!*

> Buddy, aka Earl Hewson, was the bachelor son of Albert and Lola Hewson and was born the same year as my uncle, Johnny Hansen. He lived a couple miles east of the Hansen homestead, closest to Johnny's farm.
>
> Larry Hansen told me that shortly before his father, Johnny died, he told about having a dream about pigs at Buddy's. After reading this in my manuscript, Larry realized it wasn't imagination but based on something that Johnny actually remembered.

*Dear E and C* (I wondered if it was possible that Clara and Evelyn were staying together)

*We didn't get to Regina a week ago last Saturday, or did we? We were either going to Regina or Radville ...? The roads are opening up, so we'll go someplace soon. It snowed too much. JHH.* [Johnny's initials]

*Edith and Vic* [who had been married Oct 26, 1940, and were now living in Birch Hills, SK] *will most likely come down for Christmas. Edith might teach in Dravland School.*

[Edith did, in fact, teach for the balance of that 1941- 42 schoolyear.]

*Yours truly,*
*Lloyd*
*Per Johnny*


July 16, 1942

Dear Clara,

*Seeing the typewriter is down, I will write you a letter on it. Suppose you know that Evelyn and Lillian* [Johnson, their former neighbours and friends] *were out here about a week ago. Anna went along back with them for a couple of weeks. They are either going to go for her with the car next week or she will come home on the train, next Tuesday, she sure was glad to go.*

*Well, the big wedding is over and James and Gladys* [Torkelson] *have already come home from their honeymoon.* [James had been best man for Edith and Victor's wedding.]

*We had a hailstorm here on last Saturday evening. The school land was hailed out almost completely, as I said two years ago, but this year we didn't get any hail. Near Bromhead and Torquay, it was hailed out.*

*Evelyn and I went to Midale to the Bible camp from June 26 to July 3, wish we would have had a camera and taken some pictures.* [The Midale Lutheran Bible Camp, I used to go to, was situated in a unique place along the Souris River that was expropriated for the building of the Rafferty Dam in the early 90s, although according to the history of the camp, that year's [72] Bible Camp was held for

---

[72] http://www.metochos.ca/about-us/history/

the young people at Ed Torgunrud's home in Midale, with the classes held in the church. Spending a week at Bible Camp was always a high point of my summers, as I remember.]

*Thursday evening – we had a hailstorm this afternoon. There was a dark cloud in the northwest, we were in the dugout when it started to cloud up.* [What were they doing there? Most likely swimming, Larry said.] *When we got up to the house, the wind already came. We were in the warm, when it was raining and hailing, it came down about an inch of rain and lots of mostly small hail stones. The early barley was about 50%* [destroyed by the hail]. *I don't think that the wheat and the flax were hurt much, the oats were not much in the first place*

*They have the TB testing in Oungre, and the man said that that you and Myrtle should go to the General Hospital on either Wednesday or Friday to the TB clinic and have your tests done.* [Both Myrtle and Clara may have been in Regina at this time and because there had been TB in the family, everyone had to get checked out on a regular basis.]

*Nyhus said that was the first* [Calgary?] *Stampede he had been to, but it was not the last one he was to go to. He came home Monday night.*

*The weather is nice today. There are a few clouds above but not any rain clouds yet. It is fairly wet out today. Yesterday, after the rain and during the rain, the whole yard looked like a lake.*

*They are going to take a few pigs down to Hoffer this afternoon, eight of them, I guess.* [There was a stock yard east of the elevators to hold the animals prior to them being shipped out to market on the train.] *The last few days we have been fixing the fence, most of this time. We have the cows down east in the electric fence and the last few days they have come home almost every day and night, we have them down with the electric fence on? Had a Co-op fencer down there, but it was not working well, so we took it home and put that on the Eaton fencer down there instead because* [the cows] *have been breaking out of the fence.* [Because it wasn't working. Having the wires electrified was supposed to keep the cows from going through the wires and getting out of the pasture.]

*Papa has not been feeling well the last few days, but he was up this morning.*

*We sold the little white bull calf yesterday. Clarence and I took him away in the Pontiac, he could barely get into the box. We sold him to Torjus Lee.* [JB's friend who lived near Bromhead. After John and Dina retired in Estevan, Torjus was a close neighbour over on Fourth Street.]

*We shipped Windsor about two weeks ago, we got about $106.50 for him, if we would have sold him a couple of weeks earlier, we could've got about $190 for him as the price on cattle went down on fed cows, they went down about from $.14 to $.11 a pound, they are to go down still more.* [When you live on a farm you are always aware of the price for the various commodities that are being raised. The daily market reports on the radio are listened to faithfully.]

*We finished summer fallowing a few days ago; we had almost 2 hours of work left and the drawbar on the Case tractor broke right off. We tried to get a new one, but could not get one, so we got it welded. It is still holding.* [Summer fallow is land that was not planted with a crop that year and is only cultivated lightly to keep growth of weeds under control.]

*They have been canning strawberries- the last few weeks they have canned about 50 quarts. Now they are all from the strawberry patch down by the trees, they have also canned a lot with rhubarb.*

*Lloyd and I might get a bike.* [The bike that Palmer wrote about in June 1937 must be worn out by now.] *There is a bicycle down in Torquay, at Tenold's, the place where it was on Sunday. I hope it is still there later.*

*Papa said that if you could not get a job, you should come home, if you could get a ride with somebody.* [This indicates she had finished her training course.] *I have been typing enough now, so I will close.*

*Your brother,*
*Palmer*

# HOMESTEADING IN THE LAST BEST WEST

\*\*\*

So now, this next letter is included, not because it is about me, but that it shows Grandpa's reaction to the younger generations' childrearing ideas. I probably witnessed most of the grandchildren after me getting that 'Pony Ride' on grandpa's foot. The words and music, for the Norwegian action song Grandpa would sing to his grandchildren as babies or toddlers, is on page 29 of my first book.

As I reviewed John's emigration/immigration story, I realized that my birth date was exactly forty years from the day John boarded the ship to begin his homesteading journey. So that made the highlights of what he accomplished within that space of time even more significant.

Letter from my mother, Edith to her sister, my aunt Clara:

*July 1943*

*I reread your letter, to know what you want to know. I didn't even realize it was your turn to write. Figured I owed everybody letters.*

*Better start on the most important subject first– Our Darling Wee baby. Only she's not so wee. She measured 24 ½ inches a week or so ago. But she hasn't done so well on her weight. She hadn't regained her birthweight of 9 lbs. 14 oz. last Saturday when I took her to the doctor. So, I had to give her extra feedings from the bottle, she seems to have gained, but I'll find out when I take her to the hospital for weighing tomorrow. Yes, she has lots of black or dark hair and blue eyes. She has a nice pink complexion with fine regular features. Since her nose turns up, Mrs. Boe [Edith and Vic's neighbour] thought she looked like you. But most say she looks like a Melby, Harold or her daddy. When the doctor saw her last week he said, 'What a nice baby she is!' So, when a professional man like him could take time to notice such things, we think so anyway.*

*She has been smiling now over a week, all this time she is very bright, and she follows you with her eyes, she is very active and strong for her age in spite of the fact I've been starving her by not having enough to feed her.*

[She gave up on breastfeeding.] *She was really good, sleeps all night but she can cry too. We don't believe in picking her up every time she fusses a bit.*

*Victor doesn't believe in spoiling her anyway. He gets so disgusted with everybody that comes along and must pick her up every time she cries, that's the grandparents for you. Every time, papa thought we must be hard hearted to let the poor thing cry and papa sure liked to hold her and was the proudest grandpa.*

*They* [John and Dina] *spent about 10 days of a pleasant visit up here.* [This was at the end of the west coast trip they took after Dina's operation John described in the next section of his memoir, where he talked about all the trips they had taken.]

*They came on Saturday. Sunday, we were invited to Elsie and Willy's for dinner. Monday, they rested and visited with* [my paternal grandmother], *Mrs. Melby* [Mr. Melby had passed away the previous fall, about the time Edith discovered that she was pregnant with me], *Tuesday, Mrs. Melby and I went with them to Lokkens. Wednesday and Thursday, we visited Saxhaugs, and Walter Hanson's. It was Ladies Aid there Thursday.* (These were all people originally from the south of the province.) *Mr. Saxhaug was in the hospital with a kind of flu so that's why we went to Kinistino and visited Helen, too* [his daughter, who was married to Art Kurtz]. *The folks had had lunch there, on their way out.*

*The rest of the time, they spent quietly here in town, so their day was restful. Papa got so interested in reading a book we have here, that he could scarcely leave it for two days, when he finished it. Now you don't often see him sit down to read, and mind you, he had even forgotten his glasses at home.*

*Monday evening, before they left, we had Elaine baptized at the church with a very nice service and several people out. I carried her myself. Elsie and Willie were sponsors, she was very good all through the service – looked the minister in the face all the time he was baptizing her, even had a smile for him. She looked very nice dressed in her knitted coat outside her lace and ribbon-trimmed dress and tatted cap and slippers, wish I could get a picture of her.*

The tatted cap and slippers mentioned above were a gift from Edith's Outlook College friend, Margaret Gjesdal Tomtene, who was now one of her Birch Hills' neighbours.

After my Aunt Evelyn's death, I received the baby doll that had been a gift to my aunt Evelyn from my mother, Edith, in the early thirties. I had Lys Fischer, my doll maker friend, fully restore the body of this doll, including replacing the worn out "mama" mechanism. Then, she made an ivory baptismal dress for the doll, finishing off the outfit with the tatted cap and slippers. As I finished this story, I hadn't realized until I read this letter, I recreated the picture she didn't get so I could display this special handcrafted heirloom.

*Evelyn's doll photographed with several other heirloom items. A mat, braided from strips of old overalls, used to lay inside the entrance from the garage into Evelyn's house, formerly my grandparent's home. She asked me if I wanted it when she moved to her apartment. The fabric used on the mat's outer row had frayed badly as it was not as wear resistant as the inner rows, so a binding of denim fabric gave this old mat a new lease on life. The other item here was one of my grandmother's patchwork quilts, also called crazy quilt, the ultimate in recycling odd bits of leftover fabric in random patterns.*

*This evening, we're having a regular electrical storm and how's the rain ever pouring. We could easily get us something like this after the terrific heat of the last few days. This should help both gardens and crops.*

*Our garden is good considering the lateness of planting and the weeds that have grown faster than we can keep them down. I haven't been able to do much in it myself yet and it has been very busy, besides being sick with a sore throat the last few days.*

*I'm not as strong yet as I should be, next week I'm to have a check-up from the doctor. He advised me taking sun baths.*

*Today, I had to wash baby clothes and thought I'd do a few other pieces inside, so I wouldn't have so much for Monday, but I almost thought I had to leave the whole business. This heat has probably made me feel weaker the last days. Mrs. Melby has taken the heavier part of my washing in her machine.*

*Alice* [Victor's niece, who had been helping] *went home two weeks ago. Did some gardening and housecleaning while here. But then, after those things were done away with, I think I could do these everyday jobs myself. It's so much nicer being alone anyway, it's a lot less work.*

*Phyllis, Alice, Dorothy went with Lerseths to Christopher Lake yesterday.* [Probably to Bible Camp, as Lerseth was the Lutheran pastor then.] *Walter* [their father] *had dinner with us after he had delivered them. Viola L spent most of the day with me, too, while waiting for an appointment with the doctor.*

*The blanket was very nice. Will look through my knitting book now.*

*Palmer is overseas now.* [My father's brother who was in the army had just finished his training]

*Love,*
*Edith, Vic and Elaine*

John and Dina's visit with Edith and Vic came at the end of the next trip he related in his memoir.

## Chapter 20
# TRAVELS AND TRIALS

*1943 West Coast trip by train*

When I was growing up, our family wasn't much into giving hugs.

Grandma Dina appeared to be a buxom woman but the reality was, that twice in her life, she had had breast cancer. Her curves were provided by prosthetics. One time, grandma stayed overnight at our place and was going be sleeping with me. As she removed the corset that she always wore, to put on her night gown, her flat scarred chest was a jarring image.

Perhaps, this was a reason why I don't remember getting a hug from my grandmother. Though I must say, I still knew she was a warm caring person. They just never talked about "such things."

***

Once again, John's memoir reported Dina faced another round of surgery for breast cancer. *In the spring of 1943, Dina had an operation in Regina and again we saw how God, through the doctors and nurses, gave us reason to thank and praise him. Dina was weak after her operation and hadn't seen her brothers in Seattle for years, so we decided to take a trip there.*

*When in Weyburn, I bought three quarters of land for a neighbour, Ole Saxhaug, the same day we took the train to the coast. In June, we paid $2000 and the* *rest to be paid in October. Crop was good and we got half of it so it was done quite easily.* [By investing in land even after retirement John was kept busy driving out to the farm to look after his interests.] [73] For some reason, I thought this may have been the farm the Victor Melby family bought when we moved south from Birch Hills in 1946.

***

*Dina slept in a lower berth, and I had the upper. When we got to the mountains and went around curves, I was really afraid. We could look straight up, and the mountain was so steep.*

*First, we went to Helmer Nelsons in New Westminster, BC and Dina stayed there.* Nelsons were Hansen friends from the Bromhead area in the early years. When they returned to live in Estevan their yards were kitty-corner across the back alley from each other. Their adopted daughter, Alma was like family, for she always made a fuss, even over the next generation of Hansen great-grandchildren.

*Then, I went to Castlegar to visit* [my cousin] *Christian Naess and again went through the mountains with my skin intact. I sure was dreading the way back, but as it is written, 'As your days, so shall your strength be*[74]*.' We made it back.*

*We took the train to Seattle and made our home at* [Dina's brother and sister-in-law] *Jens and Matilda's and visited many relatives and friends.*

*There was so much to see, and it was great to see a bit more of God's creation but, I for one, would never exchange Saskatchewan's Golden Fields and the best government, I know of, for any other place in the world.*

***

*We went by ferry from Seattle to Victoria. We met Clarence at the border, he was in the army then.*

---

73    **The Saga of Souris Valley R.M No.7**, p.108, John B Hansen story written by Edith Melby.

74    Deuteronomy 33:25 is the origin of this phrase.

*Four Saskatchewan boys in Basic Training:*
*Woodrow Martinson, Clarence Hansen, Harald Tangjerd and Harvey Fossum*

Harald joined the Hansen family on his marriage to Clara in 1949 and when Clarence married Vera Tedford October 1951, Harvey Fossum, became his brother-in-law for his wife, Blanche was Vera's sister.

A Canadian Pacific Railway postcard, Clarence wrote in Norwegian and mailed to Edith and Victor at Birch Hills must not have got posted until he made it back home, as it is stamped Hoffer, Sask. Feb 15, 1944. "*God dag. Takk for sist.* 'God dag,' of course is 'good day.' The sentence,[75] '*Takk for sist*' is difficult to translate, it is something we Norwegians say when we meet each other after some time, it means 'Thank you for the last time we saw each other.'

*Yes, it's me, here I am again. I'm on my way home for a couple of weeks. Will be home tomorrow pm. Will be back in Nanaimo on the 2nd of March. Hello Clara! Having a good time?* [She must have been visiting Edith and Vic at the time?] *We are at Lake Louise now, the town not the lake. Hope the weather is good when I get there. The weather was swell at Nanaimo. Let's hear from you. Love, Clarence*

---

75   I emailed a picture of what Clarence had written in Norwegian to my second cousin Bjorg and Hermod Monsen in Norway and I thank them for the translation and the explanation of this phrase I've heard dozens of times.

\*\*\*

John continued: *It's really a marvel how the tide comes in and goes out. We took a guided tour by wagon, pulled by four horses where they describe all the places we passed.*

*We took the boat back to New Westminster. We were soon rocked to sleep and landed in the morning. Helmer Nelson met us, and we stayed a few days with them, visiting and sightseeing.*

*We stopped at Banff several days, we went to the Banff Hot Springs and really enjoyed that. I think God gave us this hot water to relax and heal our body and we thank him. We went on a tour up in the mountains and the bus driver told us about everything and especially, about a Norske who had been such a good skier. He showed us different kinds of wild animals. We even saw where the water divides, some running east and some west.* (The Great Divide)

*On our trip home, when the train stopped in Calgary, we got a taxi and went to see an old neighbour. It was great to see him, as we had parted in an unfriendly manner. After 5 to 6 weeks, we came back to Weyburn.*

\*\*\*

*We had bought a quarter of land, 40 miles north of our farm and dropped in to see the fellow who had rented the land next to ours. A fellow travelling around, who owned this land, wanted me to buy the three quarters, 480 acres. I said I had to go home and see how the crop was. When I told Bernard, he said, "Buy it." So, we drove back.*

*We paid $1500 and were to get all the crop. And we got home and checked all the crops, they were beautiful, especially the hundred acres of flax on that land. This gave us an extraordinary start, so we had the land paid for, in 1944. It appears God had opened His almighty hand over us since 1938.*

\*\*\*

*Perhaps I got too involved in these worldly things, John* laments, remembering another battle he experienced

# HOMESTEADING IN THE LAST BEST WEST

*On the morning of October 28, 1944, I took Clarence to Weyburn, after his army leave* [which had allowed him] *to come home and help in the harvest.*

*That evening, Johnny took a load of grain to town. When there was no one to go for the cows, I put the saddle on Billy and went galloping after the cows. I saw the cattle, east of Hoffer, and set off in full speed, after them. Billy fell in a badger hole and all I remember is, I stood beside Billy, and the strap was broken on the saddle. I don't remember how I got back to the Hoffer Road, but I met Johnny there with the truck. Johnny went for the cattle, and I drove the truck home.*

*Dina didn't notice anything wrong with me, until I asked, 'Is the hayloft door open?'*

*Later, when I went into the bedroom, I looked out and asked, 'Are the grain bins full?'*

*Since they had recently been filled, Dina knew something was wrong with me. After that, she put me to bed with ice cold cloths on my head and I came around.* [He likely had a concussion from falling off the horse.]

***

*Not long after this, I began to itch and burn almost all over my body. I asked the doctor in Estevan, when I went to see him and asked, 'Do you think you could cure me.'*

*He said 'yes,' and I said, 'If you can cure me, I'll give you $10.'*

*He gave me a salve to smear on and then was told to bathe it off after a few hours and then smear on some more. I had a hotel room with bath, so I smeared on and washed off until at last, I fainted in the bathtub.*

*When I came to, I lay for a while in my room, but since I was to take the train* [back] *to Hoffer, someone helped me to the station and from what I heard, from a neighbour lady, I had fainted in the station, and I had thought I was on the train. I remember, one man threw his coat on the floor, and I fell on it.*

*When I got to Hoffer, I went to a friend's home to rest, until one of my sons came for me. Oh, yes, there were some long nights and days when I burned and itched all over.*

*One morning, I told Dina, 'We had better get ready to go someplace, since I'm practically burned up.' We went to Estevan by train and went to the same doctor. When I showed him my raw arms and legs, he said, 'You must go to a specialist.'*

*I also went to Mrs. Jennings, our chiropractor, who had always been so good to help us. But she said the same, so we went to Regina.*

*We got there rather late at night and there was no room at the hotel. We went to a Mrs. Canning, who advertised some rooms in her home. She gave us both rooms and meals. She phoned the skin specialist but, since I couldn't get in the next morning, I got her to call a chiropractor, named Johnson, and went to see him. He sat me on a machine which had hands like a clock, and it stopped on the word, eczema. He used his fingers good on my body and he said he could cure me.*

*I saw the skin specialist the next day. He said "You have eczema and have to go to the hospital. There are 40 kinds of eczema, so we don't know if we can heal you, but I'll do my best."*

*Dina thought, perhaps it best not to go to the hospital, but I should go to the chiropractor. The first two nights I walked in circles near the bed when it burned so frightfully, but towards morning, my thoughts went from my terrible sins to my Saviour's pain on the cross when he paid for my sins with His precious blood.*

*And yes, wonder of wonders He helped me again! Dina was with me all this time. During the day we took in various [evangelistic] meetings and got quite well acquainted. In four weeks, I was better and didn't have one red spot on my whole body.*

\*\*\*

Close to the end of his memoir, as he recorded it, JB looked at the changes that have taken place in farming and he tells how he passed on what he had built up to the next generation.

## Chapter 21
# LOOKING AT CHANGES

*Comparing 1910 and 1950*

It is almost unbelievable how much things have changed in the last few years, not least of all farm machinery. Now you can get machines for every possible thing-yes, even for picking stones, but it naturally takes so much more money. Yes it's almost impossible unless you have a lot of good land and get good prices.

It is now better that we have towns nearer the farm. [However, that is no longer the case.] I remember in the beginning we had to haul grain 20 to 30 miles, a bit later 18 miles. Then I was up at 3 o'clock and started the stove, was out to feed and harness the horses and then have a bit of breakfast and so I was usually in Colgate with the load at 9 o'clock when the stores opened, and I could sell right away and get loaded up with coal or other materials and be ready to leave for home as soon as I'd eaten dinner. It was good to get home and to bed in good time.

The next day I had to haul manure from the barn and get everything ready with hay and feed so it would be easy for Dina to take care of the cows and horses. Then I loaded another load by hand in decent time [as I had to get to bed, for I then had] *to be up again at three. At that time, I hauled 50 to 60 bushels* [in a wagon] *with two horses. Now we haul 200 to 300 bushels in big trucks and have only 3 to 4 miles to travel.* [This was in the 1950s and the changes since that period have been almost as remarkable.]

*Threshing was almost an overwhelming thing, as it took so many men. Before, it took 10 to 15 men; now, our four sons do all the work themselves with 3 to 4 times as much to go over. They can just drive the big combine, which both cuts and threshes at once* [called straight combining] *and the grain goes into a big tank holding 40 to 50 bushels, and when it is full, the grain empties by a bottom opening, into a truck that comes alongside the combine. That person drives it to the granary or bin, where in just minutes, 200 to 300 bushels is transferred* (by means of an auger) *to the bin and then back for another full load.*

*Now, they use swathers that are 12 to 16 feet wide to cut the crop, so it can dry on the stubble in swaths that are 2 to 3 feet wide. When* [it is] *dry they come with the combine with a 6-foot pickup.*

***

*I had for many years given the boys pay for their work, crop off their own land to pay for the land, which I think, gave them the incentive and interest in farming.*

*In the spring of 1945, I sold some land to each of the four oldest boys for a reasonable price and one quarter of the crop was to go to me for the land payment. In a few years, the land was paid for and was theirs. I sold them the machinery, horses and cattle reasonably. Yes, God, you've given them good crops for years and very good understanding of it all, and a real desire to work and so we thank you, God, once more.*

***

I had questions, about the various kinds of tractors they used (so I could identify them on pictures correctly) Answers were provided by my cousin, Larry Hansen, who elaborated further about later developments of the Hansen Brother's farming operation.

*The Avery (tractor) would be in the early 1920s. They would have used the Avery and the Mogul into the mid 1930's. Then they went to a 15-30 McCormick-Deering (an early International Harvester tractor). Then in early 1940s a D Case tractor, then an LA Case tractor, then a WD9 IHC (International Harvester), They had 2 of these for years, then 600 IHC, then 1960 830 John Deere, at the same time, a 560 Cockshutt tractor. Then 5010 John Deere, then 5020 John Deere.*

*Then in 1972, since they all had sons getting older and they had enough money in the Hansen Brothers account* [they decided] *to split up and start farming their own land separately. (They each had their own land all the time they farmed but owned all the machinery in a 4-way split - at the end of the year they would get a portion of the 4-way split to live on and farm into the next year.)*

*\*\*\**

And as John is easing into retirement, he says- *In the summer of 1945 Dina and I left for Manitou Beach, when I was red with eczema again. We bought a cabin, called 'Wee Blue Inn,' north of the lake, a good house with five rooms for $1200.*

I didn't remember Grandpa's cabin, though this photo indicates I visited there that summer for by November of that year my second brother, Palmer was born.

*Elaine and Dennis sitting in the special chair. 1945*

*Lloyd and Grandpa on the pier*

*When you're in the water, they say you can't sink because the water is so salty. Lay on your back with arms and legs stretched out and read a newspaper. Here I've learned to float and I'm 68 years old.*

*Friends from Hoffer, Mr. and Mrs. Hans Lund were at the lake too. He also had a skin ailment. He was feeling better and wanted to go home with us. The day before he got worse, went to the hospital where he passed away after a few days. Yes, we predict, but God rules!*

# HOMESTEADING IN THE LAST BEST WEST

***

*Since Watrous was near Saskatoon, we went there, and because our children wanted an education, we bought a good-sized house in the nicest part of the city and paid half, with the rest to be paid later; the papers were all in order. We got a letter from the lady saying she needed the house, we heard she sold it for $1800 more than the price she asked of us.*

*But, God does everything for our best. It was much better for us to build in Estevan, where we were near our children and friends.*

*After I sold to the boys in 1945, we gave each of our children $500. The four boys on the farm got $2000 off on the machinery they bought from me. For each of the others, I bought them a $500 bond to help them with something for a rainy day.*

*I thank God for our children and all the blessings he gives us, especially salvation in Jesus Christ and eternal life.*

And now John is preparing to build his retirement home in Estevan.

December 1946 family photo. My mother, Edith, is very pregnant as my brother Allen was born in early March of 1947. Back Row L-r Johnny, Anna, Lloyd, Clara, Myrtle, Bernhard, Edith, Clarence Front Row –Evelyn, John, Dina, Palmer.
A rare studio photo of myself with my brothers, Dennis and Palmer as toddlers, was likely taken then.

# Chapter 22
# RETIREMENT

In the fall of 1945, we left for Estevan, and we were blessed to get a lot, quite cheap, in a good district. The first winter, we lived with J.E. Marken, and started to buy materials for our house. I bought lumber and piled it near Elling Ellingson's house, so it would dry out.

I had to go in the car, both east and west, to haul in many of the necessary things for our house. Since it was shortly after the war, there were shortages in all lines.

Our 24' x 26' house has a full basement, a 10' x 20' veranda on front and a 26 x 10 washroom at the back. The stairway goes from the back porch leading to three second story bedrooms. This way, our two single daughters can have rooms to keep their belongings.

*JB Hansen's home in retirement*

*I bought the lot in 1945 for $50. Now in 1956, $3000 is paid for the poor ones. I saved a lot by buying and hauling our own lumber before the* [wartime] *price controls were removed.*

*Our house cost $7000, but I think it could sell for $15,000 now, as everything has gone up in price. We have a good home now in our old days and hope to sell it to Evelyn, our single daughter, so it can continue to be our family home.*

\*\*\*

I moved part of this next section of the memoir back to the Thirties chapter, but this is where he had it in his memoir, as he looked back on trips he had taken in his life. But now John's story moves forward with the trips taken after retirement.

\*\*\*

*Our next long trip was in 1949, to New Era, Michigan, where I had landed as a newcomer in 1903, and left there in 1904 so I had been away 45 years. It had changed a lot, and* [John's] *aunt and uncle had died. We stayed most of the time with* [my cousin] *Carrie Pierce and visited around from there.* John had been there for Carrie's wedding to Nathan Pierce in 1901, Nathan had passed away in 1943

*I was so glad to meet Victor Monsen, the one who first gave me work in America. He and his sons were alone on a nice farm. In the 45 years I had been away, it had changed a lot. When my youngest cousin came home, we had a good laugh. He took my hand and said "Pro" like I had blurted out, 46 years before, when I thought the young fellows drove unlawfully fast.*

*It was interesting to drive the car right onto the ferry that took us over Lake Michigan, both ways. We visited in and around Fertile, Minnesota on the way to Michigan.*

*In 1955, Lloyd took Myrtle, Evelyn, Dina and I to the 75th anniversary of Little Norway Church at Fertile Minnesota. Dina and all her brothers had been confirmed there. We made our home at Oscar Johnson's* [Dina's cousin,

Nora] *and had a real nice visit with the 10 Johnson brothers —three, on each of two large farms, were doing well at farming.* [My mother said the Johnson brothers were known as a one family baseball team.]

*On our way home, we stopped at Hitterdal and visited Henry Brekke, who wasn't that well now. He wouldn't be able to carry a grown man in his arms anymore, like he did when the truck got stuck in the slough, years before. When we were at Henry Brekke's, we read in the newspaper that there had been a terrible cyclone, southwest of Fargo, so we went by there, and it was terrible, and at the same time, interesting to see the destruction. From Hitterdal, we went to Detroit Lakes, and when we didn't find George Brekke home, we went by the place where I had hauled milk to Fargo.*

John didn't mention, probably because he stayed home, that in the fall of 1958 my uncle Lloyd drove his mother, Dina and her brother, Adolph, as well as my aunt Evelyn and my mother, Edith to Seattle for the celebration of Jens and Mathilda Gronvold's 50th wedding anniversary.

\*\*\*

*There were many special times visiting at our Grandma and Grandpa's, or them visiting us. On this occasion they visited at the Melby farm, a picture was taken for it was Dina's 70th birthday, Oct 2, 1957.*

*Attention from our aunties made everyday fun. This old 15-30 IHC was the first tractor used at the Melby farm when my father, Victor started farming. L-R Palmer, Elaine, Dennis (behind), Allen, Evelyn and Clara 1948*

Visiting back and forth between the various families happened frequently as the majority lived within a three-mile radius. Any time one of the uncles dropped by or came to visit was a special occasion. And our maiden Aunts, Myrtle and Evelyn graced each family with a visit for at least a day or two out of their 10-day school holiday at Christmas or Easter, and don't forget the summer. The Hansen nieces and nephews did get a lot of special treatment and encouragement from their single aunts. This extended down to them remembering the next generation of their grandnieces and nephews on birthdays and other occasions.

*Then, there were the times, like this day at the Melby farm, when a number of the family were together on a Sunday afternoon and here the men were inspecting the Melby's new combine. Harald Tangjerd took the picture.*
*In front of combine - Norman Melby, Carl and Lorne Tangjerd, Larry Hansen*
*On tractor- Gerald, Allen, Dennis, Palmer Melby*
*Behind tractor- Adolph Gronvold, Victor Melby, Lloyd Hansen, Johnny Hansen, Yoel Bergman (a neighbor, about a mile south, we had been doing custom harvesting for)*

And I cannot forget the combined family Christmas Eve celebrations that started in the early fifties and continued until the late 1970s. The earliest list I found among Edith's things was 1964 in an envelope labelled 'Hansen Christmas menus', though I especially remembered one from the early fifties when all the grandchildren posed with Grandma and Grandpa displaying the gift they had given us. By combining resources with Myrtle and Evelyn, we always got a significant gift from our grandparents when we were younger. One year I received a very beautiful doll, when I particularly remembered thinking that at nine years of age, I'm too old for a gift like this.

We were expected to perform something for our grandparents-such as a song or a verse we had learned for our Sunday School Program or School Christmas Concert.

> Everyone contributed to the menu and each family took a turn to host the affair. The last ones I remembered attending were at Lloyd and Lauretta's in 1969, the year before I got married and 1976, the year before my father, Victor died.

\*\*\*

JB Hansen's accomplishments were documented in the local history book,[76]

*Dad kept excellent records and loved to write. He was always interested in community affairs, Dravland School has already been mentioned. In the earlier years, he was on the board of the Grain Growers Association at Tribune. He also served on the Wheat Pool, rural telephone company and the Cooperative Commonwealth Federation (CCF) from its beginning. After he moved to Estevan, he was among the founders of the Estevan Credit Union, for which he was honoured by getting #1 as his account number. He was a charter member of the Lutheran Church in the Hoffer area, on which board he served in various capacities all through the years. Later, he was on the building committee for constructing the new Lutheran Church on 2nd Street in Estevan.*

*By investing in land, even after retirement, he was kept busy driving out to the farm to look after his interests. The garden became his hobby in the summer months.*

*In later years, his hearing gradually declined, until he lived in a world of his own.*

\*\*\*

John ends by saying, *Evelyn was so good to us, as she stayed with us in Estevan and taught school.*

*We enjoyed our stay in Estevan, as we had so many good friends, a good church- Trinity Lutheran, and a wonderful family- 10 children and lots of*

---

76   **The Saga of Souris Valley R.M. No. 7,** from the John B. Hansen Story. Page 108 written by Edith Melby

*grandchildren.* [There were 17 grandchildren in 1956 and 11 more were added, the count as of February 2021 of the descendants of John and Dina Hansen is 186.]

Other remembrances are recorded in my first book, **The Princess Doll's Scrapbook**. Now an occasion of Golden celebration brings this story to a close.

# ELAINE MELBY AYRE

*May 1956 JB Hansen Family photo*
Back row L-R Elaine Melby, Palmer, Victor Melby, Clarence, Harald Tangjerd, Bernhard, Evelyn, Olaf Hagen, Johnny, Lloyd. Middle row-L-R Vivian Tangjerd, Edith, Vera holding Garnet and Clifford to her left, Clara Tangjerd holding Carl, Grandpa and Grandma holding the twins, Janice and Kandis Hagen, Anna Hagen holding Wayne, Evelyn with Rhonda, Myrtle with Donna Hagen beside her, Seated on the floor- my brothers- Dennis, Palmer, Allen, Gerald, Norman Melby, Lorne Tangjerd and Larry Hansen)

CHAPTER 23

# THE JUBILEE

*1905-1955*
*50th Anniversary of Saskatchewan becoming a province*
*Community celebration*
*Hoffer, Saskatchewan*

For weeks, we had been practicing songs to present for the concert on the celebration day for Saskatchewan's Golden Jubilee. To help us prepare, our teacher Mrs. Ghan, who was originally Esther Hoffer, took advantage of, the 'Saskatchewan Sings' School Broadcast that came on the radio every Friday afternoon. We listened, learned and sang the several different songs celebrating our province. The songs praised our "Land of the rolling plain, sunlit and free, fruitful with golden grain, boundless to see…"

In particular, "Saskatchewan Hymn" was written and composed by Neil Harris for this School Music Broadcast Series- 'Saskatchewan Sings' for the Saskatchewan Department of Education and the Saskatchewan Golden Jubilee Committee, 1955.[77]

---

77    The Saskatchewan Hymn appears to have been published under government direction which would make the song subject to crown copyright. Crown copyright means the government owned the original rights, but they expired 50 years after publication. Since the booklet was published in 1955, Saskatchewan Hymn would have entered the public domain in 2005. Information from the Saskatchewan Archives in an email, Fall 2017

*Saskatchewan Hymn* by Neil Harris from the *Student Song Book for the School Music Broadcast Series Saskatchewan Sings*, Saskatchewan Department of Education and the Saskatchewan Golden Jubilee Committee, 1955.

\*\*\*

Finally, the selected date dawned and that perfect June day was one of the highlights of Hoffer's short history. People came from far and wide to celebrate and honour the pioneers from the Hoffer area who started their journeys in life on a homestead in the early years of Saskatchewan becoming a province. The Community Hall had been refinished inside as a tribute to the pioneers. Items from the olden days, on display both inside and out, helped the old to recall and the young to be reminded, of the many changes that had happened in the 50 years we were celebrating.

Those in my age group lead the respected guests of honour to their places at the celebration banquet honoring the district pioneers. My brother, Dennis led my grandmother while I ushered my grandfather in the parade of pioneers.

\*\*\*

*Grandmother's brothers [the Gronvold families] who had come out from western Washington State, for the Saskatchewan celebration, met with all the aunts and uncles and cousins of the Hansen family at the Clarence Hansen Farm the next day.*

\*\*\*

When I started putting ideas for this story together the movie, Camelot was on. It struck me that the words of those Saskatchewan songs we learned, described a different kind of special place. It was real not imaginary, garnered from assets and resources my grandfather and countless others like him had been faithful to develop through all their years of homesteading. And despite no end of difficulties; John had faced-the quirks of the seasons, grasshoppers, dust storms, prairie fires, blizzards, hailstorms, poor economic conditions, troubling health issues, spiritual battles yet this 'Last Best West' became his most congenial spot.

Though I knew that Grandpa ultimately looked forward to his heavenly home, in his life on this earth, Grandpa had found and been part of crafting his Camelot. It was no imaginary place: It was Saskatchewan!

\*\*\*

After a short final illness, John passed away, Jan 27, 1965. Evelyn continued to stay with Dina. I lived there for two years, too, from fall 1966 to summer of 1968, while I held a teaching position at the Collegiate. Dina passed away March 29, 1968. Evelyn lived there until apartment life at Trinity Tower, where my mother, Edith, was living, appealed to her more than living on her own.

And here's Grandfather's last word: *"Yes, we have much to thank God for- a good home while we are here in this veil of tears, but let us remember our heavenly home, which our Saviour, Jesus Christ has prepared for us and invites us to."*

*'Come unto me, all you that labour and are heavy laden, and I will give you rest.' Isn't it wonderful to look forward to the rest, which has been prepared for all those who have received grace to see themselves as sinners, and accept the invitation to become Christians?*

\*\*\*

# Afterword

*We want to dedicate this, their history to the best parents in the world, Mom and Dad —John and Dina Hansen. Dad lived to be 87 and Mom 81.....Dad always enjoyed farming. He, Mom and all of us children worked hard and we enjoyed it on the farm. We each had work to do and were an important part of it.*

*Dad enjoyed writing until the very last and wrote this story of his life in Norwegian. Edith and I had fun translating it, as he mixed English and Norwegian. It was quite funny in places.*

*The church and their many friends were very important to them, both on the farm and in Estevan.*

*"Thanks, Mom and Dad, for your Christian devotion and example to us as children and to all who loved you.*

*Clara Lillian Tangjerd 1991*

# Acknowledgements

A recent note from my friend, Jean stated that if I were to write about the medical mash unit my husband and I experienced in the past several years, nobody would believe all the twists and turns we have gone through. But here I am, needing this book to be done and I must bite the bullet and acknowledge those who have helped me make this book a reality. I want to offer thanks to family and friends, new and old for their moral and prayer support.

First, I must thank God for bringing us to Peace River, where he has led us through this valley of suffering. There were many special "God moments" as this book all came together. *Soli Deo Gloria!* To God be the Glory

I am thankful to be part of the blessing of this Hansen clan. I am thankful for the support of my husband, Gary and my sons, Will, Nate and Jed who could not understand why or how the reworking of this manuscript took so long. I love you all.

I am grateful my Aunt Clara's dream of publishing my grandfather's story will become a reality. Having this in the back of my mind did encourage me to continue, for there were countless times I thought the story was finished and it wasn't. I often despaired ever completing it. But now, I'm thankful, it is finished!

I want to acknowledge the following:

My mother's little notebooks provided a glimpse into the summers of 1933 and '34, and other of her little notes provided interesting history bites throughout the book.

My mother's photo albums provided many informal pictures in this book. She was handy with a basic box camera.

My cousin Jan Erisman gave me her mother's (my aunt) Clara Tangjerd's photo album as well as several family letters from the forties. That album provided the delightful double exposure photograph used for the cover of the book. The picture is of my uncle Johnny and represents looking back as well as looking forward- one theme for this book.

I want to honour the memory of two online friends, Phillip Ramstad and Barb Raaen who assisted on my first book. I certainly missed the editorial assistance Phillip provided for **The Princess Dolls Scrapbook**. And Barb helped me find the needed references from the **Plentywood Cookbook** for this book.

When I realized I needed an editorial evaluation, I remembered that I had dealt with Margo Dill through a blog called Women on Writing (WOW) as well as Editor 911 and that I had an offer on an editorial evaluation of a manuscript, so I thank Margo Dill for her assistance in finding ways to improve this manuscript.

To my friend, Wilma Bjorndalen, for ongoing support and friendship through the years and for providing the special words she wrote for the back cover.

To my cousin Larry Hansen, acknowledged as our resident Hansen Family Historian in the previous book; for pictures provided and many farming questions about the early days answered, especially remembrances that our uncle Bernhard had shared with him. I thank him for reading over the first and second proof of the manuscript and pointing out a few items that I had missed or made a mistake in identifying, and for mentioning a few priceless little stories that were added on this last proof.

I want to thank my friend, Marlene Atkinson for her Photoshop skills used in sharpening up a good number of the photographs that were used in this book.

To my new friend Linda Harriman, for inviting me to that writing conference and for your comments used on the back cover. Good Luck on your writing journey.

**The Saga of Souris Valley, RM#7,** which tells the stories of the families from that area from 1906-1976, was an ongoing resource, as was the updated **The Continuing Saga of Souris Valley, Saskatchewan Centennial Edition** published 2005

And thank you to several people for reading various stages of the manuscript, who shall remain nameless, in case I leave someone out.

And finally, to you, dear reader, in these days when it's more important than ever to understand the past, I hope and pray that my interpretation will aid you towards that end.

*Soli Deo Gloria!*

## About the Author

As the oldest daughter of the oldest daughter of an early homesteading family from the Rural Municipality of Souris Valley No. 7 in southeastern Saskatchewan, Elaine was always interested in the history of her immigrant grandparents. The gift of her grandmother's damaged porcelain doll inspired research to discover that "Doll's Story," the inspiration for her book, **The Princess Doll's Scrapbook**, published in 2015, which related her maternal families' emigration and immigration stories. Then her favorite aunt's dream, inspired Elaine to write this story based on her grandfather, JB Hansen's (translated, handwritten, photocopied) 82-page homesteading memoir. Elaine's 30-year career as a Home Economics teacher began in southern Saskatchewan, then to Fort Simpson, NWT where she met and married her husband, Gary then to Calgary and several surrounding rural areas- Turner Valley, Water Valley and Linden, where they raised three sons as well as Canadian horses for almost 20 years. She has two grandchildren.

After retirement from teaching, she continued to work part-time as a cook at a senior's lodge in Linden, "retiring" after 17 years when they moved to Peace River, where her husband and son operate a safety training business. She enjoys northern living in that scenic river valley, particularly music, gardening, cooking, crafts, reading and volunteering. She is thankful for recovery from a 2018 major surgery to remove a gastrointestinal tumor. This reality made getting this 'history written down before it's too late' even more significant.

Printed in Canada